Mathematical Sciences Research Institute
Publications

17

Mathematical Sciences Research Institute Publications

Volume 1	Freed and Uhlenbeck: Instantons and Four-Manifolds, Second Edition
Volume 2	Chern (ed.): Seminar on Nonlinear Partial Differential Equations
Volume 3	Lepowsky, Mandelstam, and Singer (eds.): Vertex Operators in Mathematics and Physics
Volume 4	Kac (ed.): Infinite Dimensional Groups with Applications
Volume 5	Blackadar: K-Theory for Operator Algebras
Volume 6	Moore (ed.): Group Representations, Ergodic Theory, Operator Algebras, and Mathematical Physics
Volume 7	Chorin and Majda (eds.): Wave Motion: Theory, Modelling, and Computation
Volume 8	Gersten (ed.): Essays in Group Theory
Volume 9	Moore and Schochet: Global Analysis on Foliated Spaces
Volume 10	Drasin, Earle, Gehring, Kra, and Marden (eds.): Holomorphic Functions and Moduli I
Volume 11	Drasin, Earle, Gehring, Kra, and Marden (eds.): Holomorphic Functions and Moduli II
Volume 12	Ni, Peletier, and Serrin (eds.): Nonlinear Diffusion Equations and their Equilibrium States I
Volume 13	Ni, Peletier, and Serrin (eds.): Nonlinear Diffusion Equations and their Equilibrium States II
Volume 14	Goodman, de la Harpe, and Jones: Coxeter Graphs and Towers of Algebras
Volume 15	Hochster, Huneke, and Sally (eds.): Commutative Algebra
Volume 16	Ihara, Ribet, and Serre (eds.): Galois Groups over $\mathbb{Q}$
Volume 17	Concus, Finn, and Hoffman (eds.): Geometric Analysis and Computer Graphics

P. Concus R. Finn D. A. Hoffman
Editors

Geometric Analysis and Computer Graphics

Proceedings of a Workshop held May 23–25, 1988

With 60 Illustrations, 30 in Full Color

Springer-Verlag
New York Berlin Heidelberg London
Paris Tokyo Hong Kong Barcelona

Paul Concus
Lawrence Berkeley Laboratory
and Department of Mathematics
University of California at Berkeley
Berkeley, CA 94720
USA

Robert Finn
Department of Mathematics
Stanford University
Stanford, CA 94305
USA

David A. Hoffman
Department of Mathematics
University of Massachusetts
Amherst, MA 01003
USA

Mathematics Subject Classification: 49-06, 49F20, 53-06, 53A10, 58Exx, 68U05, 76-06, 76B45, 82A57

Library of Congress Cataloging-in-Publication Data
Geometric analysis and computer graphics : proceedings of a workshop held May 23-25, 1988 / Paul Concus, Robert Finn, David A. Hoffman, editors.
p. cm. — (Mathematical Sciences Research Institute publications ; 17)
The workshop was sponsored by the Mathematical Sciences Research Institute in Berkeley, California.
Includes bibliographical references.
1. Geometry, Differential—Data processing—Congresses.
2. Computer graphics—Congresses. 3. Calculus of variations—Data processing—Congresses. I. Concus, Paul, 1933- . II. Finn, Robert, 1922-
III. Hoffman, David A., 1944- . IV. Mathematical Sciences Research Institute (Berkeley, Calif.) V. Series.
QA641.G4 1990
516.3′6′028566—dc20 90-41813

Printed on acid-free paper.

Softcover reprint of the hardcover 1st edition 1991

Camera-ready text prepared by the Mathematical Sciences Research Institute using PC T_EX.

9 8 7 6 5 4 3 2 1

ISBN-13: 978-1-4613-9713-7 e-ISBN-13: 978-1-4613-9711-3
DOI: 10.1007/978-1-4613-9711-3

Preface

This volume derives from a workshop on differential geometry, calculus of variations, and computer graphics at the Mathematical Sciences Research Institute in Berkeley, May 23–25, 1988. The meeting was structured around principal lectures given by F. Almgren, M. Callahan, J. Ericksen, G. Francis, R. Gulliver, P. Hanrahan, J. Kajiya, K. Polthier, J. Sethian, I. Sterling, E. L. Thomas, and T. Vogel. The divergent backgrounds of these and the many other participants, as reflected in their lectures at the meeting and in their papers presented here, testify to the unifying element of the workshop's central theme.

Any such meeting is ultimately dependent for its success on the interest and motivation of its participants. In this respect the present gathering was especially fortunate. The depth and range of the new developments presented in the lectures and also in informal discussion point to scientific and technological frontiers being crossed with impressive speed. The present volume is offered as a permanent record for those who were present, and also with a view toward making the material available to a wider audience than were able to attend.

We wish to express our appreciation to Irving Kaplansky, Director of MSRI, for his dedicated, personal role in making the workshop a reality, and to the MSRI and Lawrence Berkeley Laboratory staff for their expert assistance on the many details of the arrangements. The workshop received generous financial support from MSRI and LBL, and thereby from the National Science Foundation and the Department of Energy, which we gratefully acknowledge.

We are indebted to Alvy Ray Smith for joining us on the organizing committee.

Paul Concus (Chairman)
Robert Finn
David Hoffman

Contents

Preface v

Color Plates following page 108

F. J. Almgren, Jr.

Multi-functions Mod ν 1

A. Baldes and O. Wohlrab

Computer Graphics of Solutions of the Generalized Monge-Ampère Equation 19

T. F. Banchoff

Computer Graphics Tools for Rendering Algebraic Surfaces and for Geometry of Order 31

J. R. Bellare, J. A. MacDonald, P. P. Bergstrom, L. E. Scriven, and H. T. Davis

A Low Cost Animation System Applied to Ray Tracing in Liquid Crystals 39

M. J. Callahan

From Sketches to Equations to Pictures: Minimal Surfaces and Computer Graphics 49

P. Concus and R. Finn

Nonuniqueness and Uniqueness of Capillary Surfaces 63

J. L. Ericksen

Static Theory of Nematic Liquid Crystals 65

G. K. Francis

The Etruscan Venus 67

R. Gulliver

Harmonic Mappings, Symmetry, and Computer Graphics 79

W. E. Johnston, D. W. Robertson, D. E. Hall, J. Huang, F. Renema, M. Rible, and J. A. Sethian

Video Based Scientific Visualization 89

R. Kusner

Bubbles, Conservation Laws, and Balanced Diagrams 103

D. Langbein and U. Hornung

Liquid Menisci in Polyhedral Containers 109

M. L. Lapidus

Can One Hear the Shape of a Fractal Drum? Partial Resolution of the Weyl-Berry Conjecture 119

A. Linnér

Steepest Descent as a Tool to Find Critical Points of $\int k^2$ Defined on Curves in the Plane with Arbitrary Types of Boundary Conditions . . . 127

K. Polthier

Geometric Data for Triply Periodic Minimal Surfaces in Spaces of Constant Curvature 139

A. H. Schoen

Embedded Triply-Periodic Minimal Surfaces and Related Soap Film Experiments 147

J. A. Sethian

The Collapse of a Dumbbell Moving Under Its Mean Curvature 159

A. R. Smith

Geometry Versus Imaging: Extended Abstract 169

I. Sterling

Constant Mean Curvature Tori 175

D. Terzopoulos

On Deformable Models 181

E. L. Thomas, D. M. Anderson, D. C. Martin, J. T. Hoffman, and D. Hoffman

Periodic Area Minimizing Surfaces in Microstructural Science 193

T. I. Vogel

Types of Instability for the Trapped Drop Problem with Equal Contact Angles 195

Multi-functions Mod ν

FREDERICK J. ALMGREN, JR.

Abstract. We extend the theory of current valued multi-functions to multi-functions mod ν.

§1. Introduction.

This note is an introduction to and advertisement for current valued multi-functions. It is also an announcement and preliminary exposition of a corresponding new theory of multi-functions mod ν. Such multi-functions provide a novel perspective on various important problems in geometry and the calculus of variations in the context of geometric measure theory. They are indispensable in proofs of several basic theorems in geometric measure theory [**5**][**9**] and have provided a setting for both original and alternative proofs of other pivotal results [**2**][**15**]. They additionally are a device for computational geometry and its associated graphics [**3**][**7**] (see also [**8**]). The first use of such functions in the context of geometric measure theory was in [**1**]. The basic reference for current valued multi-functions is [**2**] (where they are called multiple-valued functions); expository accounts appear in [**3**][**7**]. Applications in complex analysis are set forth in [**4**][**12**][**16**]. One important example of a multi-function arises when one regards the roots of a (real or complex) polynomial as a function of its coefficients; see, for example, [**10**, 4.3.12]. The general theory of current slicing appears in [**10**, 4.3]; [**13**] is an illustrated introduction to geometric measure theory in general.

We here set forth multi-functions in the context of polyhedral chains with coefficients either in the integers or in the integers modulo ν. The extension to general rectifiable currents or flat chains mod ν is then relatively straightforward for those knowledgeable in the subtleties of geometric measure theory and perhaps irrelevant for the development of discrete algorithms in the geometric calculus of variations.

Supported in part by NSF grant DMS 84-0129-AO2

Associated with piecewise affine mappings $f, g: \mathbf{R}^n \to \mathbf{R}^N$ are induced chain mappings

$$f_\sharp, g_\sharp: \mathbf{IP}_*(\mathbf{R}^n) \to \mathbf{IP}_*(\mathbf{R}^N)$$

of degree 0 between chain complexes of integral polyhedral chains. In particular these mappings respect the additive structure of $\mathbf{IP}_*(\mathbf{R}^n)$. Since $\mathbf{IP}_*(\mathbf{R}^N)$ is also a group under addition we can define an addition of such chain mappings, e.g.

$$(zf_\sharp + wg_\sharp): \mathbf{IP}_*(\mathbf{R}^n) \to \mathbf{IP}_*(\mathbf{R}^N), \quad (zf_\sharp + wg_\sharp)(T) = z(f_\sharp T) + w(g_\sharp T),$$

whenever $z, w, \in \mathbf{Z}$ and $T \in \mathbf{IP}_k(\mathbf{R}^n)$. Caution: $(f_\sharp + g_\sharp)$ does not equal $(f+g)_\sharp$ because this latter addition uses addition in $\mathbf{R}^N$ rather than in $\mathbf{IP}_*(\mathbf{R}^N)$. It turns out that the mapping $(zf_\sharp + wg_\sharp)$ is determined by the its action on single point masses $[\![p]\!]$. The multi-function F associated with this sum is

$$F: \mathbf{R}^n \to \mathbf{IP}_0(\mathbf{R}^n), \quad F(p) = (zf_\sharp + wg_\sharp)[\![p]\!] = z[\![f(p)]\!] + w[\![g(p)]\!], \quad p \in \mathbf{R}^n.$$

In order to induce a chain mapping on $\mathbf{IP}_*(\mathbf{R}^n)$ it is not necessary that our F come from such f's and g's. Indeed, general Lipschitz continuous mappings $\mathbf{R}^n \to \mathbf{IP}_0(\mathbf{R}^n)$, locally of bounded mass, also naturally induce chain mappings on $\mathbf{IP}_*(\mathbf{R}^n)$. I do not know the extent to which general Lipschitz multi-functions can be approximated by sums of "single valued" maps. Multi-functions were studied extensively in [**2**]. As indicated above we will review some of these notions in the present context of polyhedral chains and piecewise affine multi-functions and show that they remain valid for polyhedral chains with coefficients in the integers modulo ν when the F's take values in zero dimensional polyhedral chains mod ν

§2. Simplicial decompositions and piecewise affine mappings.

(1) By a 0 *simplex* in $\mathbf{R}^n$ we mean a point in $\mathbf{R}^n$. By a k *simplex* in $\mathbf{R}^n$ we mean the convex hull Δ of some collection of $k+1$ points $p_0, p_1, \ldots, p_k$ in $\mathbf{R}^n$ which do not lie in any $(k-1)$ dimensional affine subspace of $\mathbf{R}^n$. The points $p_0, p_1, \ldots, p_k$ are called the *vertices* of Δ.

(2) When we say that $\mathbf{SX}_* = \mathbf{SX}_0 \cup \mathbf{SX}_1 \cup \ldots \cup \mathbf{SX}_n$ is a *simplicial decomposition* of $\mathbf{R}^n$ we mean that

(i) $\mathbf{SX}_0$ is a discrete set of points in $\mathbf{R}^n$ (necessarily countably infinite and unbounded).

(ii) $\mathbf{SX}_n$ is a family of n simplexes Δ in $\mathbf{R}^n$ whose union is $\mathbf{R}^n$ such that the vertices of each such simplex Δ are exactly those members of $\mathbf{SX}_0$ lying within Δ.

(iii) For $k = 1, \ldots, n-1$ each $\mathbf{SX}_k$ is the family of all k simplexes Δ_k in $\mathbf{R}^n$ whose vertices are a subset of the vertices of some member Δ_n of $\mathbf{SX}_n$.

(3) $\mathcal{H}^k$ denotes Hausdorff's k dimensional measure in $\mathbf{R}^n$. $\mathcal{H}^k(\Delta_k)$ agrees with another other reasonable definition of the k dimensional area of a k simplex Δ_k.

(4) A function $f\colon \mathbf{R}^n \to \mathbf{R}^N$ is called *affine on A* provided the restriction of f to A is also the restriction to A of some affine mapping $\mathbf{R}^n \to \mathbf{R}^N$. A function $f\colon \mathbf{R}^n \to \mathbf{R}^N$ is called *piecewise affine* provided there is some simplicial decomposition $\mathbf{SX}_*$ of $\mathbf{R}^n$ such that f is affine on each n simplex in $\mathbf{SX}_n$. Piecewise affine functions are continuous.

(5) It is sometimes useful to factor a mapping through its graph space. With this in mind, we here fix

$$\Sigma\colon \mathbf{R}^n \times \mathbf{R}^N \to \mathbf{R}^n, \qquad \Pi\colon \mathbf{R}^n \times \mathbf{R}^N \to \mathbf{R}^N$$

as projections on the factors indicated and define

$$1_{\mathbf{R}^n} \bowtie f\colon \mathbf{R}^n \to \mathbf{R}^n \times \mathbf{R}^N, \qquad (1_{\mathbf{R}^n} \bowtie f)(x) = (x, f(x))$$

whenever f maps $\mathbf{R}^n$ to $\mathbf{R}^N$. Clearly

$$\Sigma \circ (1_{\mathbf{R}^n} \bowtie f) = 1_{\mathbf{R}^n} \quad \text{and} \quad \Pi \circ (1_{\mathbf{R}^n} \bowtie f) = f.$$

The point of such factorizations is that $(1_{\mathbf{R}^n} \bowtie f)$ is one-to-one while Π is infinitely differentiable.

§3. Common refinements of simplicial decompositions.

The following proposition is useful when one wishes to make definitions based on simplicial decompositions.

PROPOSITION.

(1) *Suppose $\mathcal{S}$ is any finite collection of simplexes in $\mathbf{R}^n$ of various dimensions. Then there is some simplicial decomposition $\mathbf{SX}_*$ of $\mathbf{R}^n$ such that each k dimensional member of $\mathcal{S}$ is the union of (finitely many) members of $\mathbf{SX}_k$.*

(2) *Corresponding to each collection $\mathbf{SX}_*^{(1)}, \mathbf{SX}_*^{(2)}, \ldots, \mathbf{SX}_*^{(M)}$ of simplicial decompositons of $\mathbf{R}^n$ there is some simplicial decomposition $\mathbf{SX}_*$ of $\mathbf{R}^n$ such that each member of any $\mathbf{SX}_k^{(j)}$ is a union of (necessarily finitely many) members of $\mathbf{SX}_k$.*

§4. General currents.

For our purposes a k (dimensional) *current* in $\mathbf{R}^n$ is a continuous real valued linear function on the real vector space of infinitely differentiable k forms $\varphi: \mathbf{R}^n \to \wedge^k \mathbf{R}^n$ having compact support. The Euclidean current $\mathbf{E}^n$ assigns to each n form φ the number $\int_{\mathbf{R}^n} \langle \mathbf{e}_1 \wedge \cdots \wedge \mathbf{e}_n, \varphi \rangle \, d\mathcal{L}^n$. With several exceptions (e.g. Euclidean currents), the currents T we will consider have compact support, denoted spt$T$. The boundary of a general k current T is the $k-1$ current ∂T defined by setting $\partial T(\omega) = T(d\omega)$, e.g. in geometrically reasonable cases, Stokes's theorem becomes a definition. For proper smooth mappings $f: \mathbf{R}^n \to \mathbf{R}^N$ the image current $f_\sharp T$ is defined by setting $(f_\sharp T)(\psi) = T(f^\sharp \psi)$; this implies, in particular, that $f_\sharp \circ \partial = \partial \circ f_\sharp$ since $f^\sharp \circ d = d \circ f^\sharp$.

Integration of differential forms over Lipschitz singular chains (in the sense of algebraic topology) enables one to regard such chains as currents; one obtains the same current if a chain is subdivided or if it is changed by an orientation preserving domain reparametrization. The so called *integral currents* $\mathbf{I}_*(\mathbf{R}^n)$ are a strong closure (in the space of general currents) of the currents obtained by integration over Lipschitz singular chains (see the original definition [**11**, 3.7]) and, as such, have similar combinatorial and homological properties.

§5. Oriented simplexes and polyhedral chains.

A 0 current T in $\mathbf{R}^n$ is called an *oriented* 0 *simplex* provided $T = [\![p]\!]$ for some point p in $\mathbf{R}^n$; this means $T(\varphi) = [\![p]\!](\varphi) = \varphi(p)$ for each smooth function φ (with compact support). A k current T ($k > 0$) is called an *oriented k simplex* provided there is a k simplex Δ having vertices $p_0, p_1, \ldots, p_k$ such that

$$T = [\![p_0, p_1, \ldots, p_k]\!] = \mathbf{t}(\Delta, 1, \xi).$$

This notation means that, for each smooth differential k form φ having compact support,

$$T(\varphi) = [\![p_0, p_1, \ldots, p_k]\!](\varphi) = \mathbf{t}(\Delta, 1, \xi)(\varphi) = \int_{x \in \Delta} \langle \xi(x), \varphi(x) \rangle \, d\mathcal{H}^k x;$$

here ξ is the unit simple k vector valued orientation function on Δ which assigns to each point x in Δ the k vector $\xi(x) = \eta/|\eta|$ where $\eta = (p_1 - p_0) \wedge (p_2 - p_0) \wedge \ldots \wedge (p_k - p_0)$. If $\Delta(1), \ldots, \Delta(M)$ are distinct members of some $\mathbf{SX}$ with $\bigcup_i \Delta(i) = \Delta$ then clearly

$$\mathbf{t}(\Delta, 1, \xi)(\varphi) = \sum_{i=1}^{M} \mathbf{t}(\Delta(i), 1, \xi)(\varphi),$$

i.e. as currents oriented simplexes are identified with the sum of subdivisions. The abelian group generated by all oriented k simplexes within the vector space of general k currents is called the group of *integral polyhedral k chains* in $\mathbf{R}^n$ and is here denoted $\mathbf{IP}_k(\mathbf{R}^n)$. If $T = [\![p_0, p_1, \ldots, p_k]\!] = \mathbf{t}(\Delta, 1, \xi)$ as above, then our notational conventions are illustrated by the requirements that

$$3T = 3[\![p_0, p_1, \ldots, p_k]\!] = \mathbf{t}(\Delta, 3, \xi)$$

and

$$-7\,T = -7\,[\![p_0, p_1, \ldots, p_k]\!] = 7[\![p_1, p_0, p_2, \ldots, p_k]\!] = \mathbf{t}(\Delta, 7, -\xi).$$

The middle entries in the right most expressions above, e.g. 3 and 7, are always positive since they represent a surface density. General *rectifiable k currents* are of the form $T = \mathbf{t}(S, \theta, \xi)$ which means

$$T(\varphi) = \int_{x \in S} \langle \xi(x), \varphi(x) \rangle \, \theta(x) \, d\mathcal{H}^k x$$

for each φ; here S is a k rectifiable subset of $\mathbf{R}^n$ oriented by ξ and having integer valued density function θ. The *mass* of T is the number

$$\mathbf{M}(T) = \sup\{T(\varphi) \colon \|\varphi\| \le 1\} = \int_S \theta \, d\mathcal{H}^k.$$

It turns out that the integral currents mentioned above are precisely those rectifiable currents whose current boundaries are also rectifiable.

According to Stokes's theorem

$$\partial[\![p_0, p_1, \dots, p_k]\!] = \sum_{j-0}^{k}(-1)^j[\![p_0, \dots, p_{j-1}, p_{j+1}, \dots, p_k]\!]$$

for each oriented k simplex $[\![p_0, p_1, \dots, p_k]\!]$. This current boundary operator is a homomorphism $\partial\colon \mathbf{IP}_k(\mathbf{R}^n) \to \mathbf{IP}_{k-1}(\mathbf{R}^n)$ $(k > 0)$. The collection of polyhedral chains of all dimensions forms a chain complex in the obvious way. It is useful to extend this chain complex by adding a homomorphism

$$\partial\colon \mathbf{IP}_0(\mathbf{R}^n) \to \mathbf{Z}, \quad \partial T = T(1), \quad \text{e.g. } \partial\left(\sum_i z_i[\![p_i]\!]\right) = \sum_i z_i.$$

It follows readily from Proposition 3 that for each fixed integral polyhedral k chain T in $\mathbf{IP}_k(\mathbf{R}^n)$ there exists a simplicial decomposition $\mathbf{SX}_* = \bigcup_\alpha \mathbf{SX}_\alpha$ of $\mathbf{R}^n$ such that

$$T = \sum_{i=1}^{N} \mathbf{t}(\Delta_k(i), \theta(i), \xi(i)) \qquad (\text{some } N, \Delta_k(i), \theta(i), \xi(i))$$

where $\Delta_k(1), \dots, \Delta_k(N)$ are distinct members of $\mathbf{SX}_k$. If $k \geq 1$ we can then write

$$\partial T = \sum_{j=1}^{M} \mathbf{t}(\Delta_{k-1}(j), \sigma(j), \eta(j)) \qquad (\text{some } M, \Delta_{k-1}(j), \sigma(j), \eta(j))$$

where $\Delta_{k-1}(1), \dots, \Delta_{k-1}(M)$ are distinct members of $\mathbf{SX}_{k-1}$.

§6. The cone over a polyhedral chain.

If $T = \sum_i z_i[\![p_0(i), \dots, p_k(i)]\!]$ is an integral polyhedral chain in $\mathbf{IP}_k(\mathbf{R}^n)$ $(k \leq n-1)$ and q is a point in $\mathbf{R}^n$, then the *cone over T with vertex q* by definition equals

$$[\![q]\!] \times\!\!\!\!\times T = \sum_i z_i[\![q, p_0(i), \dots, p_k(i)]\!] \in \mathbf{IP}_{k+1}(\mathbf{R}^n)$$

and a short calculation shows

$$\partial[\![q]\!] \times\!\!\!\!\times T \begin{cases} = T - [\![q]\!] \times\!\!\!\!\times \partial T & \text{if } k \geq 1 \\ = T - (\partial T)[\![q]\!] & \text{if } k = 0. \end{cases}$$

Hence $T = \partial Q$ for some Q if and only if $\partial T = 0$.

§7. Mapping polyhedral chains by piecewise affine mappings.

Let T be an oriented k simplex $[\![p_0, \dots, p_k]\!]$ in $\mathbf{IP}_k(\mathbf{R}^n)$, and suppose $f: \mathbf{R}^n \to \mathbf{R}^N$ is affine on $\operatorname{spt} T$. Then

$$f_\sharp T = f_\sharp[\![p_0, \dots, p_k]\!] = [\![f(p_0), \dots, f(p_k)]\!] \in \mathbf{IP}_k(\mathbf{R}^N);$$

equivalently

$$f_\sharp T = \Pi_\sharp \circ (1_{\mathbf{R}^n} \bowtie f)_\sharp T = \Pi_\sharp[\![(p_0, f(p_0)), \dots, (p_k, f(p_k))]\!].$$

Similarly, $f_\sharp(zT) = zf_\sharp T$ for each integer z. It is immediate that $f_\sharp \partial T = \partial f_\sharp T$. If f is piecewise affine and T is a polyhedral k chain then, in accordance with Proposition 3 and our remark in 4, there will exist a simplicial decomposition $\mathbf{SX}_*$ of $\mathbf{R}^n$ such that f is affine on each simplex of $\mathbf{SX}_*$ and

$$T = \sum_{i=1}^{M} \mathbf{t}(\Delta(i), \theta(i), \xi(i)) \quad (\text{some } M, \Delta(i), \theta(i), \xi(i))$$

where $\Delta(1), \dots, \Delta(M)$ are distinct members of $\mathbf{SX}_k$. We then set

$$f_\sharp T = \sum_{i=1}^{M} f_\sharp \mathbf{t}(\Delta(i), \theta(i), \xi(i)).$$

With this definition it is straightforward to check that

$$f_\sharp: \mathbf{IP}_*(\mathbf{R}^n) \to \mathbf{IP}_*(\mathbf{R}^N)$$

is a well defined chain mapping of degree 0, e.g. $f_\sharp$ is a dimension preserving homomorphism with $f_\sharp \circ \partial = \partial \circ f_\sharp$ whose definition is independent (on the current level) of the particular choice of $\mathbf{SX}_*$.

§8. Zero chains and piecewise affine multi-functions.

Whenever $T \in \mathbf{IP}_0(\mathbf{R}^n)$ with $\partial T = 0$ there will exist a nonnegative integer M and (not necessarily distinct) points $p_1, \dots, p_M, q_1, \dots, q_M \in \mathbf{R}^n$ such that $T = \sum_{i=1}^M [\![q_i]\!] - \sum_{j=1}^M [\![p_j]\!]$. For such T we define

$$\mathbf{G}(T) = \inf \left\{ \sum_{i=1}^{M} |q_i - p_{\sigma(i)}| : \sigma \text{ is a permutation of } \{1, \dots M\} \right\}.$$

This is equivalent to setting

$$\mathbf{G}(T) = \inf \{\mathbf{M}(Q): Q \in \mathbf{IP}_1(\mathbf{R}^n) \text{ with } \partial Q = T\};$$

the optimal Q equals $\sum_{i=1}^{M} [\![p_{\sigma(i)}, q_i]\!]$ for the right σ. For each fixed integer z_0, there is a corresponding metric $\mathbf{G}$ on $\mathbf{IP}_0(\mathbf{R}^n) \cap \{T: \partial T = z_0\}$ defined by setting $\mathbf{G}(S, T) = \mathbf{G}(S - T)$.

When we say that f is a *piecewise affine multi-function* we mean that, for some positive integers n and N and some simplicial decomposition $\mathbf{SX}_*$ of $\mathbf{R}^n$, f maps $\mathbf{R}^n$ to $\mathbf{IP}_0(\mathbf{R}^N)$ and the following is true. Associated with each n simplex Δ_n in $\mathbf{SX}_n$ there is a nonnegative integer M together with integers $z_1, \ldots, z_M$ and affine mappings $g_1, \ldots, g_M: \mathbf{R}^n \to \mathbf{R}^N$ such that

$$f(x) = \sum_{i=1}^{M} z_i [\![g_i(x)]\!] \qquad \text{for each } x \in \Delta_n.$$

If f is such a piecewise affine multi-function, then one defines the function $[\![1_{\mathbf{R}^n}]\!] \bowtie f: \mathbf{R}^n \to \mathbf{IP}_0(\mathbf{R}^n \times \mathbf{R}^N)$ by setting

$$([\![1_{\mathbf{R}^n}]\!] \bowtie f)(x) = \sum_{i=1}^{M} z_i [\![(x, g_i(x))]\!] = [\![x]\!] \times f(x)$$

for each x in Δ_n, etc. If f is a piecewise affine multi-function then it is straightforward to check the existence of some integer z_0 such that $\partial \circ f(x) = z_0$ for each $x \in \mathbf{R}^n$; furthermore, f is $\mathbf{G}$ continuous.

As an example, the function $f: \mathbf{R} \to \mathbf{IP}_0(\mathbf{R})$,

$$f(x) = \begin{cases} [\![x]\!] - [\![-x]\!] & \text{if } x > 0 \\ 0 & \text{if } x \leq 0 \end{cases}$$

is a piecewise affine multi-function.

In view of Proposition 3, the piecewise affine multi-functions from $\mathbf{R}^n$ to $\mathbf{IP}_0(\mathbf{R}^N)$ themselves form an abelian group based on the addition within $\mathbf{IP}_0(\mathbf{R}^N)$.

§9. Local representation of multi-functions.

The analysis of piecewise affine multi-functions and multi-functions mod ν is facilitated by the following proposition.

PROPOSITION. *Suppose* Δ *is a nondegenerate* k *simplex in* $\mathbf{R}^n$ $(k \geq 1)$ *and* $z_1, \ldots, z_M$ *are integers and* $g_1, \ldots, g_M: \mathbf{R}^n \to \mathbf{R}^N$ *are affine functions such that*

$$\sum_{i=1}^{M} z_i [\![g_i(x)]\!] = 0$$

$$\left[\text{resp.} \sum_{i=1}^{M} z_i [\![g_i(x)]\!] \in \nu \mathbf{IP}_0(\mathbf{R}^N) \qquad \text{for some } \nu \in \{2, 3, 4, \ldots\}\right]$$

for each $x \in \Delta$. *Then there is a partitioning of* $\{1, \ldots, M\}$ *into nonempty subsets* $W_1, \ldots, W_K$ *such that, for each* $\alpha = 1, \ldots, K$,

(1) $(g_i|\Delta) = (g_j|\Delta)$ *whenever* $i, j \in W_\alpha$;

(2) $\sum_{i \in W_\alpha} z_i = 0$ $\left[\text{resp. } \sum_{i \in W_\alpha} z_i \in \nu \mathbf{Z}\right]$.

§10. Mapping polyhedral chains by piecewise affine multi-functions.

Suppose $f: \mathbf{R}^n \to \mathbf{IP}_0(\mathbf{R}^N)$ is a piecewise affine multi-function and T is an integral polyhedral k chain in $\mathbf{R}^n$. In accordance with Proposition 3 we can find some simplicial decomposition $\mathbf{SX}_*$ of $\mathbf{R}^n$ with respect to which the following is true.

(1) $T = \sum_{i=1}^{M} \mathbf{t}(\Delta(i), \theta(i), \xi(i))$ for some M, $\Delta(i)$, $\theta(i)$, and $\xi(i)$ where $\Delta(1), \ldots, \Delta(M)$ are distinct members of $\mathbf{SX}_k$.

(2) For each $i = 1, \ldots, M$ and each $x \in \Delta(i)$,

$$f(x) = \sum_{j=1}^{J(i)} z(i, j) [\![f(i, j)(x)]\!]$$

for some nonnegative integers $J(1), \ldots, J(M)$, some integers $z(i, j)$, and some affine functions $f(i, j): \mathbf{R}^n \to \mathbf{R}^N$.

We then set

$$f_\sharp T = \sum_{i=1}^{M} \sum_{j=1}^{J(i)} z(i, j) f(i, j)_\sharp \mathbf{t}(\Delta(i), \theta(i), \xi(i)) \in \mathbf{IP}_k(\mathbf{R}^N).$$

The obvious extension of this definition defines

$$([\![1_{\mathbf{R}^n}]\!] \bowtie f)_\sharp \mathbf{E}^n \in \mathbf{IP}_{n,loc}(\mathbf{R}^n \times \mathbf{R}^N)$$

and also

$$f_\sharp \mathbf{E}^n = \Pi_\sharp \circ ([\![1_{\mathbf{R}^n}]\!] \bowtie f)_\sharp \mathbf{E}^n \in \mathbf{IP}_n(\mathbf{R}^N)$$

in case $\{x: f(x) \neq 0\}$ is bounded (such boundedness implies, in particular, that $\partial \circ f = 0$).

§11. Multi-functions induce chain mappings.

One of the basic properties of multi-functions is the following.

THEOREM. *Suppose* $f: \mathbf{R}^n \to \mathbf{IP}_0(\mathbf{R}^N)$ *is a piecewise affine multi-function. Then the induced mapping of polyhedral chains*

$$f_\sharp: \mathbf{IP}_*(\mathbf{R}^n) \to \mathbf{IP}_*(\mathbf{R}^N)$$

is a chain mapping of degree zero, e.g. $f_\sharp$ *is a dimension preserving homomorphism with* $\partial \circ f_\sharp = f_\sharp \circ \partial$.

PROOF: For example, use Proposition 3 and the factorization $f_\sharp = \Pi_\sharp \circ ([\![1_{\mathbf{R}^n}]\!] \bowtie f)_\sharp$.

§12. Slicing an integral polyhedral chain by an othogonal projection.

Suppose $(x_0, y_0), \dots, (x_n, y_n) \in \mathbf{R}^n \times \mathbf{R}^N$ are the vertices of an n simplex Δ^* in $\mathbf{R}^n \times \mathbf{R}^N$ and $x_0, \dots, x_n$ are vertices of an n simplex Δ in $\mathbf{R}^n$. Under these conditions there is a unique affine function $f: \mathbf{R}^n \to \mathbf{R}^N$ such that $f(x_i) = y_i$ for each i. By the *slice of* $T = [\![(x_0, y_0), \dots, (x_n, y_n)]\!] = \mathbf{t}(\Delta^*, 1, \xi)$ *by* Σ *at* $x \in \mathbf{R}^n \sim \partial\Delta$ we mean the integral zero chain $\langle T, \Sigma, x\rangle$ whose value at $x \in \Delta \sim \partial\Delta$ equals

$$\operatorname{sign}(\xi(x, f(x)) \bullet \mathbf{e}_1 \wedge \dots \wedge \mathbf{e}_n) \cdot [\![(x, f(x))]\!]$$

and whose value at $x \in \mathbf{R}^n \sim \Delta$ equals 0. We do not attempt to define $\langle T, \Sigma, x\rangle$ for $x \in \partial\Delta$ (Federer's treatment of slicing [**10**, 4.3] does define $\langle T, \Sigma, x\rangle$ as an appropriate real zero chain). Similarly we define

$$\langle zT, \Sigma, x\rangle = z\,\langle T, \Sigma, x\rangle \qquad \text{for each integer } z.$$

Now suppose $\Delta^*(1), \dots, \Delta^*(M)$ are n simplexes in $\mathbf{R}^n \times \mathbf{R}^N$ whose orthogonal projections $\Delta(i) = \Sigma[\Delta^*(i)]$ $(i = 1, \dots, M)$ are n simplexes in $\mathbf{R}^n$ and that

$$S = \sum_{i=1}^{M} \mathbf{t}(\Delta^*(i), \theta(i), \xi(i)) \in \mathbf{IP}_n(\mathbf{R}^n \times \mathbf{R}^N) \qquad (\text{some } M, \theta(i), \xi(i)).$$

By the slice of S by Σ at $x \in \mathbf{R}^n \sim \bigcup_{i=1}^{M} \partial\Delta(i)$ we mean the integral zero chain

$$\langle S, \Sigma, x\rangle = \sum_{i=1}^{M} \Big\langle \mathbf{t}(\Delta^*(i), \theta(i), \xi(i))\,, \Sigma\,, x \Big\rangle.$$

§13. Slicing is the inverse of multi-function mapping.

One of the central facts of multi-function theory is the following.

THEOREM. *Suppose* $\Delta^*(1), \ldots, \Delta^*(M)$ *are* n *simplexes in* $\mathbf{R}^n \times \mathbf{R}^N$ *whose projections* $\Delta(i) = \Sigma[\Delta^*(i)]$ $(i = 1, \ldots, M)$ *are* n *simplexes in* $\mathbf{R}^n$ *and that*

$$T = \sum_{i=1}^{M} \mathbf{t}(\Delta^*(i), \theta(i), \xi(i)) \in \mathbf{IP}_n(\mathbf{R}^n \times \mathbf{R}^N) \qquad (\textit{some } \theta(i), \xi(i)).$$

Set

$$g\colon \left(\mathbf{R}^n \sim \bigcup_{i=1}^{M} \partial\Delta(i)\right) \to \mathbf{IP}_0(\mathbf{R}^n \times \mathbf{R}^N), \qquad g(x) = \langle T, \Sigma, x\rangle \quad \textit{for each } x.$$

Then

(1) *For each open subset* U *of* $\mathbf{R}^n$ *a necessary and sufficient condition that* $g|U$ *be the restriction of a* $\mathbf{G}$ *continuous map* f *from* U *to* $\mathbf{IP}_0(\mathbf{R}^n \times \mathbf{R}^N)$ *is that* $\partial T \llcorner U \times \mathbf{R}^N = 0$.

(2) *If* $\partial T = 0$ *and* $f\colon \mathbf{R}^n \to \mathbf{IP}_0(\mathbf{R}^n \times \mathbf{R}^N)$ *is the continuous extension of* g, *then* f *is* $\mathbf{G}$ *Lipschitz and is the unique piecewise affine multi-function for which* $f_\sharp[\mathbf{E}^n] = T$.

REMARK: This theorem shows that integral polyhedral cycles in "general position" have a unique representation as the Lipschitz multi-function image of a Euclidean current. It is the ability of multi-functions to represent such cycles which lies at the heart of their usefulness. [**2**, Theorem 3.16] treats cycles which are not in general position. See also the multiplicity of projection estimates of [**14**].

§14. Polyhedral chains mod ν mappings, boundaries, and slices.

Associated with each integer z and with each integer $\nu \geq 2$ is the non-negative integer

$$|z|^{(\nu)} = \inf\{|w| : w - z \in \nu\mathbf{Z}\}$$

called the *ν-norm* of z. Also associated with such z and ν is a sign

$$\sigma^{(\nu)}(z) = \begin{cases} 1 & \text{if } z \notin \nu\mathbf{Z} \text{ with } |z|^{(\nu)} - z \in \nu\mathbf{Z} \\ -1 & \text{if } z \notin \nu\mathbf{Z} \text{ with } |z|^{(\nu)} - z \notin \nu\mathbf{Z} \\ 0 & \text{if } z \in \nu\mathbf{Z}. \end{cases}$$

For example, $|5|^{(7)} = 2 = |9|^{(7)}$ and $\sigma^{(7)}(5) = -1$, $\sigma^{(7)}(9) = +1$. The *polyhedral k chains* mod ν in $\mathbf{R}^n$ are by definition the quotient

$$\mathbf{IP}_k^{(\nu)}(\mathbf{R}^n) = \frac{\mathbf{IP}_k(\mathbf{R}^n)}{\nu\mathbf{IP}_k(\mathbf{R}^n)}.$$

If $T = \mathbf{t}(\Delta, \theta, \xi)$ is θ times an oriented k simplex, then the equivalence class of T is the polyhedral k chain mod ν denoted

$$T^{(\nu)} = \begin{cases} \mathbf{s}(\Delta, \omega, \eta) & \text{if } \sigma^{(\nu)}(\theta) \neq 0 \\ 0 & \text{if } \sigma^{(\nu)}(\theta) = 0; \end{cases}$$

here $\omega = |\theta|^{(\nu)}$ and $\eta = \sigma^{(\nu)}(\theta) \cdot \xi$.

The requirements that $\partial(T^{(\nu)}) = (\partial T)^{(\nu)}$ and $f_\sharp(T^{(\nu)}) = (f_\sharp T)^{(\nu)}$ for each $T \in \mathbf{IP}_k(\mathbf{R}^n)$ and each piecewise affine mapping $f: \mathbf{R}^n \to \mathbf{R}^N$ give $\mathbf{IP}_*^{(\nu)}(\mathbf{R}^n)$ the structure of a chain complex and induce chain mappings $f_\sharp: \mathbf{IP}_*^{(\nu)}(\mathbf{R}^n) \to \mathbf{IP}_*^{(\nu)}(\mathbf{R}^N)$.

Suppose $k \geq 1$ and $T \in \mathbf{IP}_k(\mathbf{R}^n)$ with $\partial T^{(\nu)} = 0$; this means $\partial T \in \nu\mathbf{IP}_k(\mathbf{R}^n)$. Set $S = T - [\![q]\!] \times\!\!\!\!\times \partial T \in \mathbf{IP}_k(\mathbf{R}^n)$ for some point q in $\mathbf{R}^n$. Then $S^{(\nu)} = T^{(\nu)}$ and $\partial S = 0$. We conclude that $T^{(\nu)} \in \mathbf{IP}_k^{(\nu)}(\mathbf{R}^n)$ *is a cycle if and only there is a cycle* $S \in \mathbf{IP}_k(\mathbf{R}^n)$ *such that* $S^{(\nu)} = T^{(\nu)}$. (The fact that this holds for general flat chains mod ν rather that only polyhedral ones is one of the major results of the new paper [**6**].) The fact that S is a boundary was noted above.

As in the integer coefficient case we define a mapping

$$\partial: \mathbf{IP}_0^{(\nu)}(\mathbf{R}^n) \to \mathbf{Z}_\nu = \frac{\mathbf{Z}}{\nu\mathbf{Z}}, \qquad \partial T^{(\nu)} = (\partial T)^{(\nu)} = [\partial T + \nu\mathbf{Z}].$$

If $T^{(\nu)} \in \mathbf{IP}_0^{(\nu)}(\mathbf{R}^n)$ with $\partial T^{(\nu)} = 0$ we can set $Q = [\![q]\!] \times\!\!\!\!\times T \in \mathbf{IP}_1(\mathbf{R}^n)$ and check that $\partial Q^{(\nu)} = T^{(\nu)}$. Hence a necessary and sufficient condition that $T^{(\nu)} \in \mathbf{IP}_0^{(\nu)}(\mathbf{R}^n)$ be a boundary is that $\partial T^{(\nu)} = 0$.

§15. Zero chains mod ν piecewise affine multi-functions mod ν and slices of a polyhedral chain mod ν by an orthogonal projection.

Whenever $T^{(\nu)} \in \mathbf{IP}_0^{(\nu)}(\mathbf{R}^n)$ with $\partial T^{(\nu)} = 0$ we set

$$\mathbf{G}(T^{(\nu)}) = \inf\left\{\mathbf{G}(S): S \in \mathbf{IP}_0(\mathbf{R}^n) \text{ with } \partial S = 0 \text{ and } S^{(\nu)} = T^{(\nu)}\right\}.$$

Whenever $R^{(\nu)}, T^{(\nu)} \in \mathbf{IP}_0(\mathbf{R}^n)$ with $\partial R^{(\nu)} = \partial T^{(\nu)}$ we similarly define the metric distance $\mathbf{G}(R^{(\nu)}, T^{(\nu)}) = \mathbf{G}(R^{(\nu)} - T^{(\nu)})$. This $\mathbf{G}$ metric is a bit more subtle than that for integer coefficients. Suppose, for example, p is the origin in $\mathbf{R}^2$, and a, b, c are points equally spaced around the unit circle in $\mathbf{R}^2$, and $T = [\![a]\!] + [\![b]\!] + [\![c]\!]$. Then $\partial T^{(3)} = 0$, and it turns out that $\mathbf{G}(T^{(\nu)}) = 3 = \mathbf{M}(Q)$ where $Q = [\![p, a]\!] + [\![p, b]\!] + [\![p, c]\!] \in \mathbf{IP}_1(\mathbf{R}^2)$ with $\partial Q = T - 3[\![p]\!] = S$ so that $(\partial Q)^{(3)} = S^{(3)} = T^{(3)}$.

When we say that f is a *piecewise affine multi-function* mod ν we mean that, for some positive integers n and N and some simplicial decomposition $\mathbf{SX}_*$ of $\mathbf{R}^n$, f maps $\mathbf{R}^n$ to $\mathbf{IP}_0^{(\nu)}(\mathbf{R}^N)$ and the following is true. Associated with each n simplex Δ_n in $\mathbf{SX}_n$ there is a nonnegative integer M and (not necessarily distinct) affine functions $g_1, \ldots, g_M$ mapping $\mathbf{R}^n$ to $\mathbf{R}^N$ such that

$$f(x) = \sum_{i=1}^{M} [\![g_i(x)]\!]^{(\nu)} \qquad \text{for each } x \in \Delta_n.$$

One can use Proposition 9 to check that each piecewise affine multi-function mod ν is $\mathbf{G}$ continuous.

If $T^{(\nu)} \in \mathbf{IP}_n^{(\nu)}(\mathbf{R}^n \times \mathbf{R}^N)$ then we define the *slice of* $T^{(\nu)}$ *by* Σ *at* $x \in \mathbf{R}^n$ by requiring

$$\left\langle T^{(\nu)}, \Sigma, x \right\rangle = \langle S, \Sigma, x \rangle^{(\nu)} \in \mathbf{IP}_0^{(\nu)}(\mathbf{R}^n \times \mathbf{R}^N)$$

whenever the right hand side of the equality exists for some S with $S^{(\nu)} = T^{(\nu)}$.

§16. Piecewise affine multi-functions mod ν are quotients of piecewise affine multi-functions.

The following theorem simplifies analysis mod ν

THEOREM. *Suppose f is a piecewise affine multi-function mod ν mapping $\mathbf{R}^n$ to $\mathbf{IP}_0^{(\nu)}(\mathbf{R}^N)$. Then there is a piecewise affine multi-function h (not mod ν mapping $\mathbf{R}^n$ to $\mathbf{IP}_0(\mathbf{R}^N)$ such that $f(x) = [h(x)]^{(\nu)}$ for each x.*

PROOF (SKETCH): It is convenient here to use the terminology of our references [**2**] and [**10**]. Suppose

$$f(x) = \sum_{i=1}^{M} [\![g_i(x)]\!]^{(\nu)} \qquad \text{for } x \in \Delta_n \in \mathbf{SX}_n$$

as in section 15 above. We then associate to Δ_n the *integral* polyhedral n chain

$$T(\Delta_n) = \sum_{i=1}^{M} ([\![1_{\mathbf{R}^n}]\!] \bowtie g_i)_\sharp \, \mathbf{t}(\Delta_n, 1, \mathbf{e}_1 \wedge \cdots \wedge \mathbf{e}_n).$$

The current

$$T = \sum \{T(\Delta): \Delta \in \mathbf{SX}_n\}$$

belongs to $\mathbf{IP}_{n,loc}(\mathbf{R}^n \times \mathbf{R}^N)$ (with the obvious interpretation) and has boundary ∂T which belongs to $\mathbf{IP}_{n-1,loc}(\mathbf{R}^n \times \mathbf{R}^N)$. We then use Proposition 9 to infer that, in fact, $\partial T \in \nu\mathbf{IP}_{n-1,loc}(\mathbf{R}^n \times \mathbf{R}^N)$ with $\operatorname{spt}\partial T \subset \bigcup\{\Delta_{n-1} \times \mathbf{R}^N: \Delta_{n-1} \in \mathbf{SX}_{n-1}\}$. We now choose and fix a point q which does not lie in any of the $n-1$ simplexes in $\mathbf{SX}_{n-1}$. For convenience of exposition we assume q is the origin.

For large numbers S we define $\sigma_S: (\mathbf{R}^n \sim \{0\}) \times \mathbf{R}^N \to \mathbf{R}^n \times \mathbf{R}^N$ by setting

$$\sigma_S(x, y) = \begin{cases} \left(S\frac{x}{|x|}, y\right) & \text{if } 0 < |x| \le S \\ (x, y) & \text{if } |x| \ge S. \end{cases}$$

Also for large numbers S we define $\tau_S: \mathbf{R} \times (\mathbf{R}^n \sim \{0\}) \times \mathbf{R}^N \to \mathbf{R}^n \times \mathbf{R}^N$ by setting

$$\tau_S(t, x, y) = \begin{cases} \left(x + \frac{x}{|x|}\min\{t, S - |x|\}, y\right) & \text{if } 0 < |x| \le S \\ (x, y) & \text{if } |x| \ge S \end{cases}$$

checking that

$$\tau_S(0, x, y) = (x, y) \quad \text{and} \quad \tau_S(S, x, y) = \sigma_S(x, y).$$

For each such S we define

$$Q_S = \tau_{S\sharp}([\![0, S]\!] \times \partial T) \in \nu\mathbf{I}_{n,loc}(\mathbf{R}^n \times \mathbf{R}^N)$$

(not **IP**) we use the homotopy formula for currents [F 4.1.9] to infer that $\sigma_{S\sharp}\partial T - \partial T = \partial Q_S$.

Suppose $0 < S_0 < S$.

(i) If $|t| \le S_0/2$ and $|x| \le S_0/2$ then $\tau_S(t, x, y) = (x + tx/|x|, y)$ which is independent of S.

(ii) If $|t| \geq S_0/2$ and $|x| \leq S_0/2$ then $\tau_S(t, x, y) = \left(x + \frac{x}{|x|}(S - |x|), y\right)$; in particular $\left|x + \frac{x}{|x|}(S - |x|)\right| \geq S_0/2$. We use these estimates (i) and (ii) to infer the existence of

$$Q = \lim_{S \to \infty} Q_S \in \nu \mathbf{I}_{n,loc}(\mathbf{R}^n \times \mathbf{R}^N)$$

with $-\partial T = \partial Q$.

Unfortunately, the Q we have constructed need not be polyhedral. To remedy this we let $\mathbf{K}$ denote the collection of unit $n+N$ cubes in $\mathbf{R}^n \times \mathbf{R}^N$ associated with the integer lattice. We then fix a rotation Θ of $\mathbf{R}^n \times \mathbf{R}^N$ such that $\dim(\Sigma \circ \Theta(V)) = n - 1$ for each coordinate $n-1$ plane V in $\mathbf{R}^n \times \mathbf{R}^N$. We apply the deformation construction [A2 1.14] to write (analogous to [A2 2.1]) $P - Q = \partial W - R$ where $P \in \nu \mathbf{IP}_{n-1,loc}(\mathbf{R}^n \times \mathbf{R}^N)$ is an integral chain of the cubical complex associated with $\Theta(\mathbf{K})$ and $R \in \nu \mathbf{IP}_{n,loc}(\mathbf{R}^n \times \mathbf{R}^N)$ since $\partial Q = -\partial T$. We set

$$T_* = T + P + R \in \nu \mathbf{IP}_{n,loc}(\mathbf{R}^n \times \mathbf{R}^N)$$

and check that

$$\partial T_* = \partial T + \partial(Q + \partial W) = \partial T - \partial T + 0 = 0$$

with $P + R \in \nu \mathbf{IP}_{n,loc}(\mathbf{R}^n \times \mathbf{R}^N)$.

We refer to Theorem 13 to let $g: \mathbf{R}^n \to \mathbf{IP}_0(\mathbf{R}^n \times \mathbf{R}^N)$ be the unique $\mathbf{G}$ continuous mapping for which

$$g(x) = \langle T_*, \Sigma, x \rangle \qquad \text{for almost every } x.$$

Finally we set $h = \Pi \circ g$.

§17. Mapping polyhedral chains mod ν by piecewise affine multi-functions mod ν.

Suppose $f: \mathbf{R}^n \to \mathbf{IP}_0^{(\nu)}(\mathbf{R}^N)$ is a piecewise affine multi-function mod ν and $T^{(\nu)} \in \mathbf{IP}_k^{(\nu)}(\mathbf{R}^n)$. We define $f_\sharp(T^{(\nu)}) = (h_\sharp T)^{(\nu)}$ where $h: \mathbf{R}^n \to \mathbf{IP}_0(\mathbf{R}^N)$ is a piecewise affine multi-function such that $h(x)^{(\nu)} = f(x)$ for each x (see Theorem 16). With this definition it is easy to check that *each piecewise affine multi-function mod ν $f: \mathbf{R}^n \to \mathbf{IP}_0^{(\nu)}(\mathbf{R}^N)$ induces a chain mapping*

$$f_\sharp: \mathbf{IP}_*^{(\nu)}(\mathbf{R}^n) \to \mathbf{IP}_*^{(\nu)}(\mathbf{R}^N)$$

of degree zero.

§18. Slicing mod ν is the inverse of multi-function mapping mod ν.

Analogous to the integer coefficient case we have.

THEOREM. *Suppose* $\Delta^*(1), \ldots, \Delta^*(M)$ *are* n *simplexes in* $\mathbf{R}^n \times \mathbf{R}^N$ *whose projections* $\Delta(i) = \Sigma[\Delta^*(i)]$ $(i = 1, \ldots, M)$ *are* n *simplexes in* $\mathbf{R}^n$ *and that*

$$S = \sum_{i=1}^{M} \mathbf{s}(\Delta^*(i), \theta(i), \xi(i)) \in \mathbf{IP}_n^{(\nu)}(\mathbf{R}^n \times \mathbf{R}^N) \qquad (\text{some } \theta(i), \xi(i)).$$

Let

$$f: \mathbf{R}^n \to \mathbf{IP}_0(\mathbf{R}^n \times \mathbf{R}^N), \qquad f(x) = \langle S, \Sigma, x \rangle$$

whenever this is defined. Then

(1) $f(x)$ *is defined for almost every* x *in* $\mathbf{R}^n$.
(2) *A necessary and sufficient condition that there exist a piecewise affine multi-function* g *mod* ν *such that* $f(x) = g(x)$ *for almost every* x *is that* $\partial S = 0$.
(3) *If* $\partial S = 0$ *and* g *is as in (2) then* $g_\sharp\left[(\mathbf{E}^n)^{(\nu)}\right] = S$.

REMARK: This theorem shows that polyhedral cycles mod ν in "general position" have a unique representation as the Lipschitz multi-function mod ν image of a Euclidean chain mod ν. This device enables one on occasion to use a function theoretic approach to problems in surface theory (such as searching for optimal geometries). In [**6**], for example, this leads to a new function theoretic proof of the compactness theorem for general flat chains mod ν analogous to that for currents set forth in [**2**].

REFERENCES

1. F. Almgren, *The homotopy groups of the integral cycle groups*, Topology **1** (1962), 257–299.
2. F. Almgren, *Deformations and multiple-valued functions*, Proc. Symposia in Pure Math. **44** (1986), 29–130.
3. F. Almgren, *Applications of multiple valued functions*, Geometric Modeling: Algorithms and New Trends, G. E. Farin, editor, SIAM, Philadelphia (1986), 43–54.
4. F. Almgren, *What can geometric measure theory do for several complex variables?*, Proceedings of the Several Complex Variables Year at the Mittag–Leffler Institute, 1987–88, (to appear).
5. F. Almgren, *$\mathbf{Q}$-valued functions minimizing Dirichlet's integral and the regularity of area minimizing rectifiable currents up to codimension two*, (multilithed notes) (1984), 1720 pp.
6. F. Almgren, *A new look at flat chains mod ν*, (in preparation).
7. F. Almgren and B. Super, *Multiple valued functions in the geometric calculus of variations*, Astérisque **118** (1984), 13–32.
8. K. Brakke, *The Evolver Program*, The Geometry Supercomputer Project (1989), (in preparation).
9. S. Chang, *Two dimensional area minimizing integral currents are classical minimal surfaces*, J. Amer. Math. Soc. **1** (1988), 699–778.
10. H. Federer, *Geometric Measure Theory*, Grundlehren math. Wiss., Springer–Verlag, New York **153** (1969), xiv + 676 pp.
11. H. Federer and W. H. Fleming, *Normal and integral currents*, Ann. of Math. **72** (1960), 458–520.
12. F. R. Harvey and H. B. Lawson, *Complex analytic geometry and measure theory*, Proc. Symposia in Pure Math. **44** (1986), 261–274.
13. F. Morgan, *Geometric Measure Theory. A Beginners Guide*, Academic Press, Inc. New York (1988), viii + 145 pp.
14. D. Nance, *The multiplicity of generic projections of n-dimensional surfaces in $\mathbf{R}^{n+k}$ $(n + k \le 4)$*, Proc. Symposia in Pure Math. **44** (1986), 320–341.
15. B. Solomon, *A new proof of the closure theorem for integral currents*, Indiana Univ. Math. J. **33** (1984), 393–418.
16. H. Whitney, *Complex Analytic Varieties*, Addison–Wesley Publishing Co., Reading, MA (1972), xii + 399 pp.

Department of Mathematics, Princeton University, Princeton, NJ 08544

Computer Graphics of Solutions of the Generalized Monge–Ampère Equation

ALFRED BALDES AND ORTWIN WOHLRAB

Abstract. Using the method of Minkowski or Alexandrov one finds simple discretizations of elliptic Monge–Ampère equations, including the equation of graphs with prescribed positive Gaussian curvature. It is shown how these discrete problems can be solved numerically, and computer graphics of the piecewise linear, convex solutions are presented.

1. Introduction.

Provided a sufficiently regular surface $\mathcal{S} \subset \mathbf{R}^3$ can be represented as a graph

$$\mathcal{S} = \mathcal{S}_z = \{(x, y, z(x, y)) : (x, y) \in \Omega\}$$

of a function $z = z(x, y)$ defined in some domain $\Omega \subset \mathbf{R}^2$, its most important intrinsic geometric invariant, the Gauss curvature, can be computed as

$$\kappa = \frac{z_{xx}z_{yy} - z_{xy}^2}{(1 + z_x^2 + z_y^2)^2}.$$

Thus, the inverse problem of trying to find a graph over Ω of prescribed Gauss curvature amounts to the solution of the Monge–Ampère equation

$$z_{xx}z_{yy} - z_{xy}^2 = \varphi(x, y)(1 + z_x^2 + z_y^2)^2. \tag{1.1}$$

Somewhat more generally we consider here the equation

$$z_{xx}z_{yy} - z_{xy}^2 = \varphi(x, y)R(z_x, z_y), \tag{1.2}$$

with $R = R(p, q)$ defined in $\mathbf{R}^2$, but immediately restrict ourselves to the elliptic case $\varphi = \varphi(x, y) > 0$ on Ω and $R = R(p, q) > 0$ in $\mathbf{R}^2$. Without further restriction $z = z(x, y)$ can therefore assumed to be convex. Because of its widespread occurrence in differential geometry the resulting Monge–Ampère equation serves as an important model in the study of the Dirichlet problem for a fully nonlinear partial differential equation of second order

$$\begin{cases} z_{xx}z_{yy} - z_{xy}^2 = \varphi(x, y)R(z_x, z_y) \text{ in } \Omega \\ z(x, y) = h(x, y) \text{ on } \partial\Omega. \end{cases} \tag{1.3}$$

In this context it is natural to assume also the domain Ω to be convex, in order not to exclude even trivial boundary data $h = h(x, y)$ right from the beginning.

Research supported in part by DOE Grant DE-FG02-86ER250125.

2. Numerical Procedure.

We want to approximate classical solutions of (1.3) by piecewise linear convex functions solving corresponding discrete problems first introduced by Minkowski [4] and later revived by Alexandrov [1].

THEOREM 1. *Let* $\Omega \subset \mathbf{R}^2$ *be a convex polygon with corners* $\underline{B} = [B_s,\ 1 \le s \le k]$ *and distinguished interior points* $\underline{A} = [A_j,\ 1 \le j \le N]$. *Then for all boundary values* $\underline{h} = [h_s,\ 1 \le s \le k]$ *and positive weights* $\underline{\mu} = [\mu^j,\ 1 \le j \le N]$ *with*

$$\sum_{j=1}^{N} \mu^j < A(R) := \int_{\mathbf{R}^2} \frac{dpdq}{R(p,q)} \tag{2.1}$$

there exists a unique convex, piecewise linear function $z = z(x,y)$ *on* Ω, *such that*

$$\begin{aligned} \omega(R,z,A_j) := \textstyle\int_{\chi_z(A_j)} \frac{dpdq}{R(p,q)} = \mu^j,\ 1 \le j \le N, \\ z(B_s) = h_s,\ 1 \le s \le k, \end{aligned} \tag{2.2}$$

and such that all interior vertices of the graph S_z *project onto* $\underline{A}$, *all boundary vertices onto* $\underline{B}$.

Here χ_z denotes the generalized gradient map with

$$\chi_z(x_\circ, y_\circ) := \left\{ \begin{array}{l} (p,q) \in \mathbf{R}^2 : z(x,y) \ge z(x_\circ,y_\circ) + p(x-x_\circ) + q(y-y_\circ) \\ \qquad\qquad\qquad\qquad \text{for all } (x,y) \in \Omega \end{array} \right\}$$

Thus $\chi_z(x_\circ, y_\circ)$ is the set of all supporting directions in $(x_\circ, y_\circ) \in \Omega$, and it is implicitly assumed that $R = R(p,q)$ is at least locally integrable.

THEOREM 2. *Let* Ω *be an arbitrary uniformly convex, bounded domain in* $\mathbf{R}^2$, $\varphi = \varphi(x,y) \in C^\alpha(\Omega)$, $R = R(p,q) \in C^\alpha(\mathbf{R}^2)$, *both positive*, $\alpha \in (0,1)$. *Assume*

$$\int_\Omega \varphi(x,y)dxdy < A(R) = \int_{\mathbf{R}^2} \frac{dpdq}{R(p,q)} \tag{2.3}$$

and $\varphi(x,y) \le \kappa d^\beta$, $R(p,q) \le C(1+p^2+q^2)^\gamma$ *with positive constants* κ, c, γ, β, $2\gamma \le 3+\beta$, $d := \mathrm{dist}_{\partial\Omega}(x,y)$. *Moreover, suppose we choose a sequence of convex polygons* Ω_m, $m \ge 1$, $\Omega_1 \subset \Omega_2 \subset \ldots$, *with* $\cup_{m=1}^\infty \Omega_m = \Omega$, *of boundary points* $\underline{B}^m = [B_s^m,\ 1 \le s \le k(m)]$, $B_s^m \in \partial\Omega_m$, *of interior points*

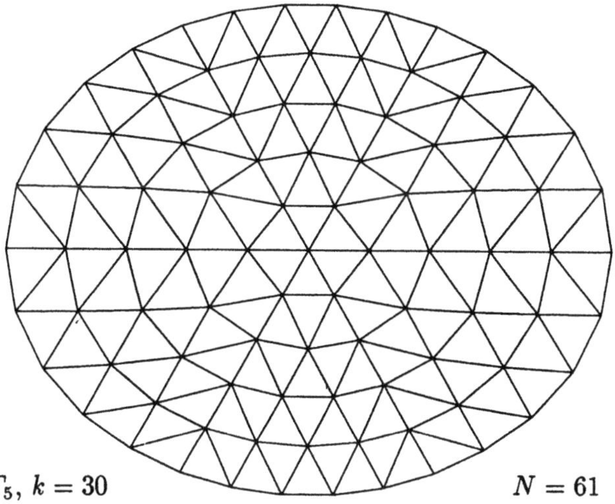

T_5, $k = 30$ $N = 61$

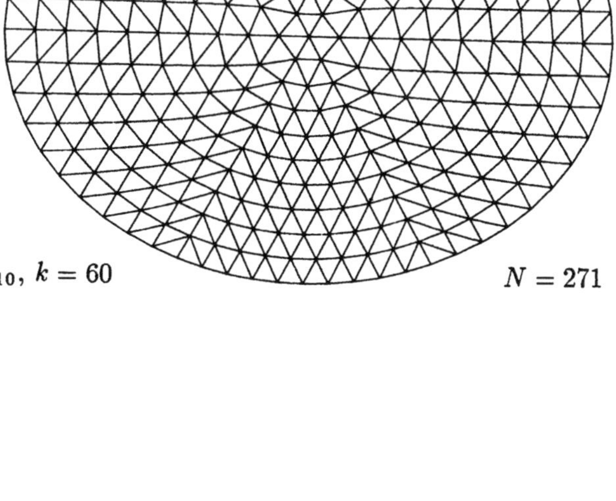

T_{10}, $k = 60$ $N = 271$

$\underline{A}^m = [A_j^m,\ 1 \le j \le N(m)]$, $A_j^m \in \Omega_m$, $\lim_{m\to\infty} k(m) = \lim_{m\to\infty} N(m) = \infty$, *and of positive weights* $\underline{\mu}_m = [\mu_m^j,\ 1 \le j \le N(m)]$, $\sum_{j=1}^{N(m)} \mu_m^j < A(R)$, *such that for every disk* $D = D_\rho(x_\circ, y_\circ)$ *with radius* $\rho > 0$ *and center* $(x_\circ, y_\circ) \in \mathbf{R}^2$

$$\lim_{m\to\infty} \sum_{A_m^j \in D} \mu_m^j = \int_{D\cap\Omega} \varphi(x,y)dxdy.$$

Then, given any sequence of boundary values $\underline{h}^m = [h_s^m,\ 1 \le s \le k(m)]$ *with piecewise linear extensions on* $\partial\Omega_m$ *converging uniformly to* $h = h(x,y)$ *on* $\partial\Omega$, *the piecewise linear, convex solutions* $z_m = z_m(x,y)$, $(x,y) \in \Omega_m$, *of the corresponding discrete problems* (2.2) *converge uniformly on* Ω *to a unique solution* $z \in C^{2,\alpha}(\Omega) \cap C^\circ(\overline{\Omega})$ *of the classical problem* (1.3).

For arbitrary Ω uniform convexity means that through every boundary point there exists a circle enclosing Ω with uniformly bounded radii.

Each function z_m being defined only on Ω_m, uniform convergence on Ω of course means that for all $\varepsilon > 0$ these exists $m_\circ$ with

$$|z_m(x,y) - z(x,y)| < \varepsilon \text{ for } m \ge m_\circ \text{ and } (x,y) \in \Omega_m.$$

Convergence of the boundary values may be defined via homeomorphisms between $\partial\Omega_m$, $m \ge 1$, and $\partial\Omega$, provided by projection along rays from a fixed origin. For a more detailed discussion of the results above we refer to [**2**] and the references therein.

Given any bounded uniformly convex $\Omega \subset \mathbf{R}^2$, we might assume its center of mass being the origin, and think of $\partial\Omega$ given in polar coordinates. Then, Ω_m, $m \ge 1$, can be taken to be the polygon determined by its boundary vertices $\underline{B}^m = [B_s^m,\ 1 \le s \le k]$ on $\partial\Omega$, where

$$B_s^m = (r(\theta_s)\cos\theta_s, r(\theta_s)\sin\theta_s), \theta_s = \frac{2\pi}{k}(s-1),$$

and we always take $k = 6m$. Then, for any boundary function $h \in C^\circ(\partial\Omega)$ we can fix $\underline{h}^m = [h_s^m,\ 1 \le s \le k]$ by setting $h_s^m = h(B_s^m)$. We chose the number of interior points in $\underline{A}^m = [A_j^m,\ 1 \le j \le N]$ as $N = 1 + 3m(m-1) = 1 + \sum_{i=1}^{m-1} 6i$ with $A_\circ^m$ the origin and the remaining $3m(m-1)$ points distributed on $(m-1)$ equidistant layers as

$$A_j^m = (\tfrac{i}{m} r(\theta_s)\cos\theta_s,\ \tfrac{i}{m} r(\theta_s)\sin\theta_s),\ 1 \le i \le m-1,$$
$$1 \le s(i) \le 6i,\ \theta_s = \pi\tfrac{s-1}{3i},\ j = 1 + 3i(i-1) + s.$$

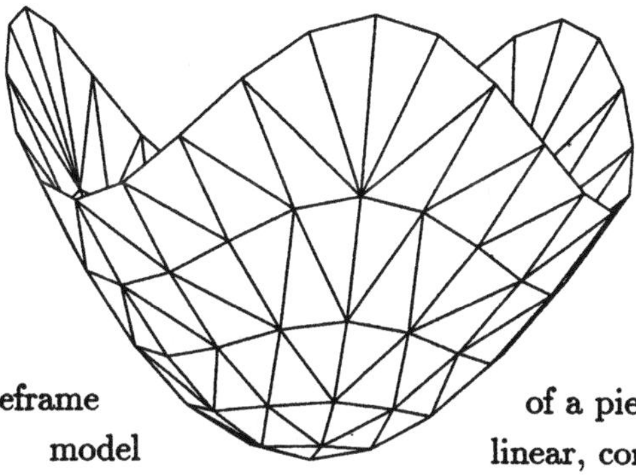

Wireframe model of a piecewise linear, convex graph

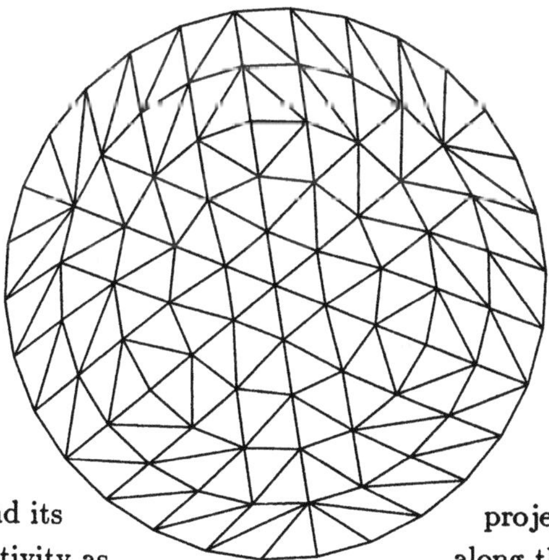

... and its connectivity as projection along the z-axis

Starting from a central hexagon it is easy to determine a regular triangulation T_m of Ω_m with interior vertices $\underline{A}^m$ and boundary vertices $\underline{B}^m$. Given a function $\varphi = \varphi(x,y) \in C^\alpha(\Omega)$, the weights $\underline{\mu}_m = [\mu_m^j,\ 1 \le j \le N]$ are taken to be

$$\mu_m^j = \frac{1}{3}\sum_{i=1}^{6} \int_{\Delta_i(A_j)} \varphi(x,y)dxdy, \text{ where } \Delta_i(A_j),\ 1 \le i \le 6,$$

are the triangles of our triangulation T_m having A_j as vertex. Then, if with some $R = R(p,q) \in C^\alpha(\mathbf{R}^2)$ inequality (2.3) and the growth conditions on φ and R in Theorem 2 are satisfied, the corresponding sequence of discrete solutions $z_m = z_m(x,y)$ will indeed converge to the desired solution of (1.3). As explained in [**2**], taking the boundary conditions into account, the function $z_m = z_m(x,y)$ is uniquely determined by its values in the vertices $z_{m,j} = z_m(A_j^m)$, $1 \le j \le N$. Thus a problem as (2.2) with Ω, z, $\underline{A}$, $\underline{\mu}$, $\underline{B}$, $\underline{h}$ replaced by Ω_m, z_m, $\underline{A}^m$, $\underline{\mu}_m$, $\underline{B}^m$, $\underline{h}^m$ respectively constitutes a nonlinear system in $N(m)$ variables, which we solved numerically by a Newton method with variable step length.

Difficulties arise from the fact, that, as necessary in each step, in order to compute the values $\omega(R, z_m, A_j)$, $1 \le j \le N$, and the derivatives with respect to $z_{m,j}$ with all vertices $(A_j^m, z_m(A_j))$ given, we have to find the unique possible connectivity satisfying the convexity condition. We have to determine the whole graph of z_m with its facets and edges as part of the convex hull of both all vertices in the interior and on the boundary. The most important use of a variable step length is to ensure that all points $(A_j^m, z_m(A_j))$ are in fact true vertices of the corresponding graph, and all sets $\chi_{z_m}(A_j)$ in gradient space have nonvoid interior. For more details we have to refer to [**2**] again.

3. Computer Graphics.

The first numerical computations of functions $z_m = z_m(x,y)$ approximating solutions of (1.3) where done on an IBM mainframe at the University of California at Santa Cruz. There we also could get reasonable wireframe and raster computer graphics, but there was no fast and immediate connection between the computing engine and the graphics device. We continued our work at the SFB 256 in Bonn, where the final version of our program has been developed on the Silicon–Graphics IRIS-3D workstation. For debugging and performance tests on our algorithms the immediate availability of even simple wireframe graphics at any given moment during program execution turned out to be extremely helpful. Because often one is not so much interested in the solution of a single particular Dirichlet problem but rather would like to study the behavior of the solution as boundary values, parameters in the equation, or even the domain are changing, our program allows the on-runtime specification of several parameters which pick the desired functions φ, R, h and the domain Ω out of one-parameter families encoded in corresponding subroutines. One more parameter controls the graphics display; you might want to see a model of the final result only, or alternatively get a glimpse of the graph S_z after each single Newton step.

As starting point of this numerical imitation of the continuation method mentioned above it is useful to have some explicitly known solutions. For Ω an ellipse,

$$\Omega = \{(x,y) \in \mathbf{R}^2 : \frac{x^2}{a^2} + y^2 < 1\},$$

and constant right hand side $c > 0$, we have

$$z(x,y) = b\left(\frac{x^2}{a^2} + y^2\right) - b, \; b = \frac{a\sqrt{c}}{2},$$

as solution of $z_{xx}z_{yy} - z_{xy}^2 = c$ with vanishing boundary data. A spherical cap yields a graph with constant Gauss curvature; thus for the unit disk $D = D_1(0)$ and constant c, $0 < c \leq 1$, we get

$$z(x,y) = \sqrt{R^2 - r^2} - \sqrt{R^2 - 1}, \; R^2 = \frac{1}{c}, \; r^2 = x^2 + y^2;$$

as solution of

$$z_{xx}z_{yy} - z_{xy}^2 = c(1 + z_x^2 + z_y^2)^2$$

with trivial boundary data again.

For example to study the one-parameter family of functions solving

$$z_{xx}z_{yy} - z_{xy}^2 = 4 \text{ in the unit circle } D,$$
$$z(x,y) = \lambda \cos\theta \text{ for } 0 \leq \theta < 2\pi, \; x = \cos\theta, \; y = \sin\theta,$$

we start with the paraboloid $z(x,y) = x^2 + y^2 - 1$ solving the problem for $\lambda = 0$; then we change the parameter step by step, always using the solution found in the step before to construct reasonable initial values for our Newton solver. The shaded images Ia, Ib, Ic show the solutions for $\lambda = 0.2$, $\lambda = 0.6$, $\lambda = 1.0$ respectively. All graphs shown in this section were computed with the choice $m = 20$, $k = 120$, $N = 1141$, and consist of $6m^2 = 2400$ triangular facets. II shows the solution of (1.3) for

$$\varphi(x,y) = 1 + x + 2y^2, \quad R(p,q) = (1 + p^2 + q^2)^{\frac{3}{2}},$$

and constant boundary data. The family corresponding to $\varphi(x,y) = 1 + \mu(x + 2y^2)$, $0 \leq \mu \leq 1$ had been considered.

Returning to the case of prescribed Gauss curvature $R(p,q) = (1 + p^2 + q^2)^2$ we must take $\beta \geq 1$ in Theorem 2. Thus, $\varphi = \varphi(x,y)$ has to vanish Lipschitz continuously on $\partial\Omega$, for us to be sure that a solution of (1.3) exists for all boundary values. Nevertheless even if this last conditions is not met, all discrete solutions $z_m = z_m(x,y)$ will still exist provided inequality (2.3) hence condition (2.1) is satisfied. The functions solving

$$z_{xx}z_{yy} - z_{xy}^2 = \tfrac{3}{4}(1 + z_x^2 + z_y^2)^2 \text{ in } D$$
$$z(x,y) = \lambda \cos 3\theta \text{ for } 0 \leq \theta < 2\pi, \; x = \cos\theta, \; y = \sin\theta$$

for $\lambda = 0.2$ and $\lambda = 0.4$ are shown by IIIa, IIIb. All graphics in this paper have been produced at the University of Massachusetts at Amherst using the VPL graphics programming environment, as described in [**3**]. The output device was the Apple LaserWriter and for color display the RasterTechnologies Model I/380. Please see Color Plates 1–3 for the color displays of Figures Ic, II, and IIIa.

After fixing a coloring of the lower hemisphere, in RGB-terms for example

$$R = -z, \; G = \tfrac{1+x}{2}, \; B = \tfrac{1+y}{2},$$
$$x^2 + y^2 + z^2 = 1, \; z \leq 0,$$

we could color each facet differently according to its unit outward normal; we used simple flat shading at this point, for Gouraud shading later resulted in less interesting images. Some specular parallel light and point light sources were added to produce a more realistic impression. IVa, IVb, IVc show solutions on the ellipse

$$\Omega = \{(x,y) \in \mathbf{R}^2 : \frac{x^2}{a^2} + y^2 < 1\}$$

with $a = \frac{3}{2}$, corresponding to

$$\varphi(x,y) = \frac{1}{2} + \mu(2 - x^2), \; R(p,q) = (1 + p^2 + q^2)^2$$
$$z(x,y) = h(\theta) = \lambda \cos 2\theta \text{ on } \partial\Omega$$

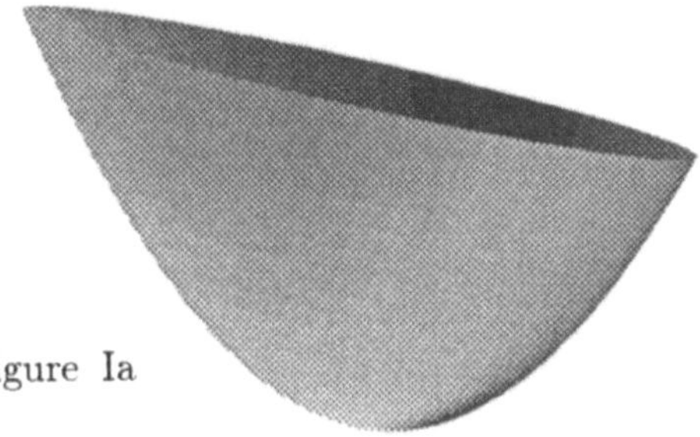

Figure Ia

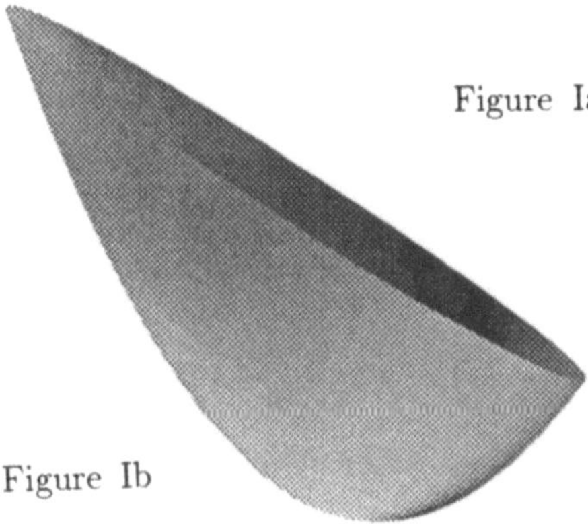

Figure Ib

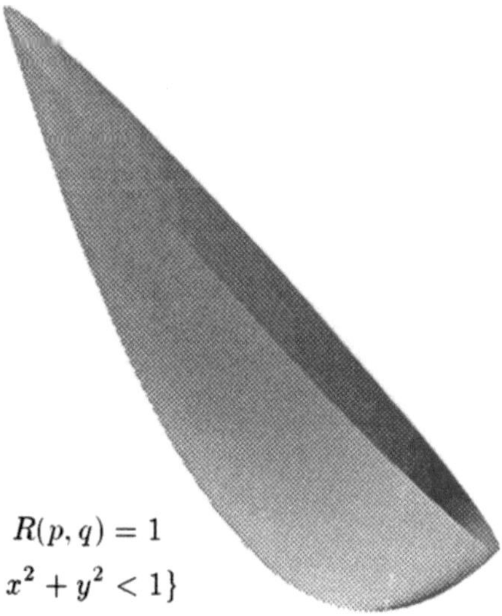

Figure Ic : $\varphi(x,y) = 4$; $R(p,q) = 1$

$\Omega = \{(x,y) : x^2 + y^2 < 1\}$

$h(\theta) = \cos\theta$

Figure II : $\varphi(x,y) = 1 + x + 2y^2$; $R(p,q) = (1+p^2+q^2)^{\frac{3}{2}}$
$\Omega = \{(x,y) : x^2 + y^2 < 1\}$
$h(\theta) = 0$

Figure IIIa : $\varphi(x,y) = 3/4$; $R(p,q) = (1+p^2+q^2)^2$
$\Omega = \{(x,y) : x^2 + y^2 < 1\}$
$h(\theta) = 1/5 \ \cos\theta$

Figure IIIb

Figure IVa

Figure IVc : $\varphi(x,y) = 1/2 + 1/8(2 - x^2)$; $R(p,q) = (1 + p^2 + q^2)^2$

$\Omega = \{(x,y) : 4x^2/9 + y^2 < 1\}$

$h(\theta) = 3/10 \ \cos 2\theta$

for $(\lambda,\mu)=(0,0)$, $(\lambda,\mu)=(0,\frac{1}{8})$, $(\lambda,\mu)=(\frac{3}{10},\frac{1}{8})$ respectively. Please see Color Plate 4 for the color display of Figure IVc. Here the deformation of a spherical cap into the graph IVa had to be computed first.

References

1. A. D. Alexandrov, *Dirichlet's problem for the equation* Det $\|u_{ij}\| = \Phi(z_1 \ldots z_n, z, x_1 \ldots x_n)$, Vestnik Leningrad Univ. Math. Mekh. Astronom **13** (1958), 5–24.
2. A. Baldes and O. Wohlrab, *Convex discrete solutions of generalized Monge–Ampère equations*, Preprint SFB 256, Bonn (to appear).
3. M. J. Callahan, D. Hoffman and J. T. Hoffman, *Computer graphics tools for the study of minimal surfaces*, Communications of the ACM 31 (1988), 648–661.
4. H. Minkowski, *Volumen and Oberfläche*, Math. Ann. **57** (1903), 447–495.

A. Baldes, Mathematisches Institut, Beringstr. 4, 53 Bonn, Fed. Rep. Germany
O. Wohlrab, Sonderforschungsbereich 256, Wegelerstr. 8, 53 Bonn, Fed. Rep. Germany

Computer Graphics Tools for Rendering Algebraic Surfaces and for Geometry of Order

THOMAS F. BANCHOFF

New developments in interactive computer graphics make it possible for mathematicians to approach old subjects in fresh ways. Occasionally the new approaches reveal additional insights into things already well understood from other viewpoints. This note give several examples of projects developed in collaboration with undergraduate students at Brown University in courses related to differential geometry. In each case, computer graphics techniques are used to investigate some geometric phenomenon, and in each case the difficulties encountered by the program reveal some feature of geometric interest.

The first example comes from an encounter with algebraic geometry which took place at the Mathematical Science Research Institute two years before this conference. In the summer of 1986, shortly before the International Congress of Mathematicians in Berkeley, Friedrich Hirzebruch was giving a series of lectures on algebraic varieties with large numbers of singular points [4]. He gave as an important motivating example a quartic surface first suggested by Čmutov. The surface is a level set of the translation hypersurface

$$F_4(x, y, z) = T_4(x) + T_4(y) + T_4(z)$$

where $T_4(x) = -8x^4 + 8x^2 - 1$ is the fourth Tschebycheff polynomial. The surface will have the symmetries of the cube, and the singularities will occur when all three coordinates are critical points of the function T_4.

The critical points of T_4 are 0 and $\pm\frac{1}{\sqrt{2}}$, and the critical values are respectively -1 and 1. The critical values of F_4 are then -3, -1, 1, and 3, containing one point, six points, twelve points, and eight points respectively. Although the equation can be interpreted as an expression in complex numbers, the singularities all occur on the corresponding real varieties so it becomes significant to consider the real loci in 3-space.

This example provided a good test of software which had been developed in student projects over the past few years at Brown University under the direction of the author. Steven Ritter and Kevin Pickhardt wrote their first implicit function renderer six years ago and the design was improved the following year by Trey Matteson. During the Summer of 1986, Edward Chang, working at the computer graphics laboratory at Brown via phone conversations with the author, was able to utilize this program to obtain rendered images of level sets of this function. These were then express-mailed to Berkeley for use as illustrations in the next talk in Hirzebruch's series (as well as in a number of subsequent lectures). See Color Plates 17a–c.

This example illustrates some other aspects of the interaction between mathematicians and computer scientists in the development of techniques of interest to both. The rendering algorithms developed at Brown, like those produced at a number of other places at approximately the same time, work well for surfaces without singularities and, predictably enough, have difficulty are precisely the points of most interest to the mathematician, the critical points where the qualitative form of the the surface changes. By providing the computer scientist with good test cases, the mathematician helps to refine the algorithms, a good symbiosis which facilitates the interaction between the disciplines.

In the case at hand, the difficulty already occurs in the two-dimensional version of the problem, namely the level curves of the function of two variables $F_4(x,y) = T_4(x) + T_4(y)$.

The critical points of $F_4(x,y)$ are a local minimum at $(0,0)$ with value -2, four local maxima at $(\pm\frac{1}{\sqrt{2}}, \pm\frac{1}{\sqrt{2}})$ with value 2, and four non-degenerate saddle points at $(\pm\frac{1}{\sqrt{2}}, 0)$ and $(0, \pm\frac{1}{\sqrt{2}})$ with value 0. A picture of the zero critical level containing the four saddle points suggests that it can be decomposed into two ellipses. Although this fact might not be so clear from a rough sketch, it is readily apparent in a computer graphics rendition of the region between levels -0.1 and 0.1, from the level curve routine in the program VECTOR, the senior project at Brown University of Rashid Ahmad in June 1988.

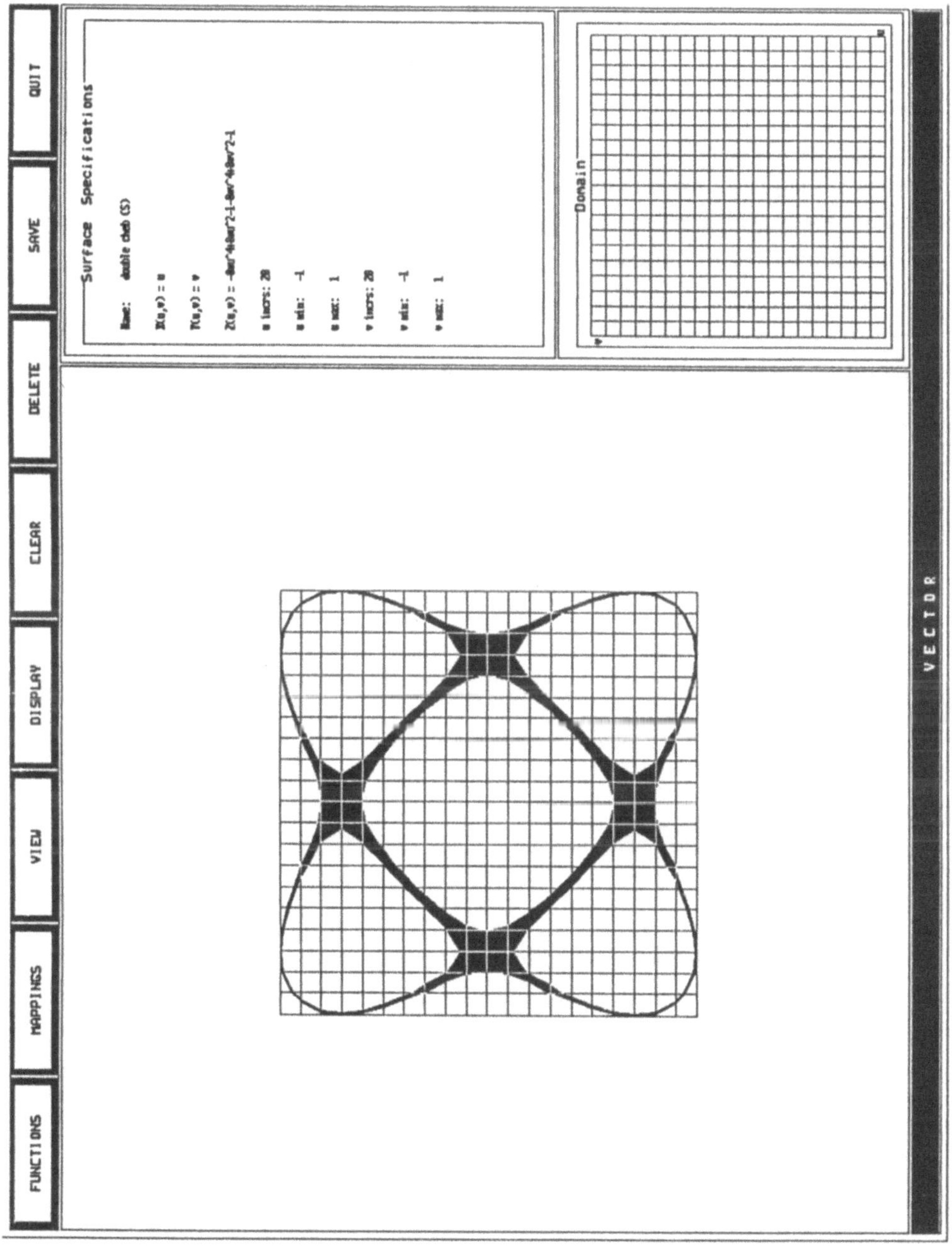
QUIT
SAVE
DELETE
CLEAR
DISPLAY
VIEW
MAPPINGS
FUNCTIONS
Surface Specifications
X(u,v) = u
Y(u,v) = v
u incrs: 20
u min: -1
u max: 1
v incrs: 20
v min: -1
v max: 1
Domain
VECTOR

The zero critical level decomposes into a product of two ellipses:

$$\begin{aligned} -8(x^4+y^4)+8(x^2+y^2)-2 \\ &= -2\left[(2x^2+2y^2-1)2-4x^2y^2\right] \\ &= -2(2x^2+2xy+2y^2-1)(2x^2-2xy+2y^2-1). \end{aligned}$$

The program used by Ahmad for displaying contour lines utilizes a straightforward approach which is suitable for a relatively small machine. For each square in a grid, the function values at the four vertices are evaluated, and depending on the configuration of these values relative to the critical value, one of three sets of level curves is drawn. The value on an edge with one vertex above and one below is computed by linear interpolation and the connections are made according to the following table:

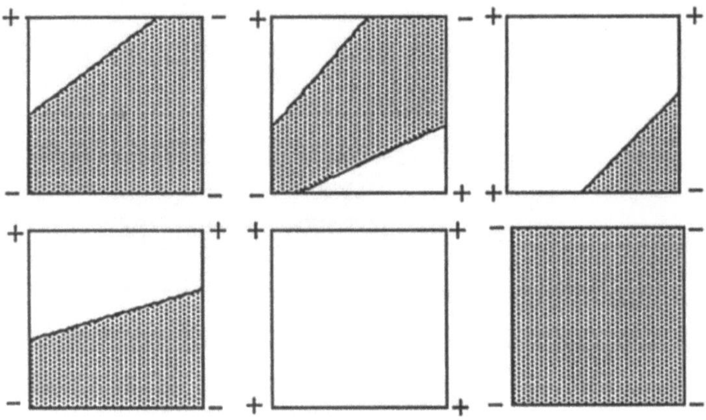

Naturally, this scheme has difficulty at precisely the points of most interest, where the four vertices have nearly the same value and each edge has one vertex above and one below the value of the level. Various improvements of the program can help to resolve the ambiguity near such points, involving selective refinements or additional tests to determine which of three possible fillings is to be preferred. There is no method which will work satisfactorily for all possible examples.

A variation of this method is the one used by Ritter and Pickhardt for rendering implicit functions in 3-space. The values of a function are calculated at the vertices of a rectangular grid, and the program them compares the configuration of signs with a collection of cubes with signed vertices. On each face of the cube, the planar algorithm determines a collection of edges. These are filled in with polygons in the interior of the cube according to the table. The resulting collection of polygons is then shipped to a rendering package which displays the level surface. Once again, the subtleties occur at those configurations where the collection of plus signs or minus signs is not connected on the cube.

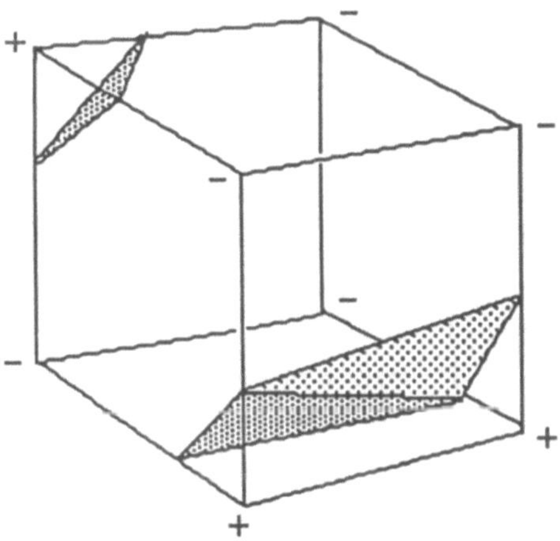

One observation that arises from the visualization process is that certain shapes are familiar from other contexts. One of the level sets for example appears in the recent volume of mathematical models from the last century ([**2**], p. 19), as a wooden model of a surface with the symmetry of a cube, as studied by Goursat [**3**]. It is difficult to construct models which possess singular points since the models tend to fall apart at conical singularities. Computer graphics representations have the virtue of not being subject to forces like gravity.

A second example in which computer graphics techniques can illustrate geometric phenomena is in the geometry of order. A surface in 3-space has geometric order four if almost any line intersects the surface in at most four points. In an article on the subject by Nicolaas Kuiper and the author, there is described an algebraic surface of the fifth degree which has geometric order four. Although much was known about the surface, it was not until the development of the implicit function rendering program that we were able to see what the object looked like.

It is more difficult to determine the geometric order of non-algebraic surfaces, and even some relatively simple cases can be deceptive. Already in the early part of this century, C. Juel raised the question of determining all tori of revolution with geometric order four [**5**]. We now have a graphics tool for investigating such phenomena, namely the programs "color-by-order" and "color-by-hit" developed by Trey Matteson as part of his senior project. The user of the color-by-order program chooses a direction and the computer indicates by different colors the points in which a line intersects the surface, yellow for order two, blue for four, violet for six and so forth. The color-by-hit program colors red the first point of the surface encountered by a ray parallel to the given direction, then yellow for the second hit, green for the third and so on. There is a certain amount of indeterminacy where a ray parallel to the chosen direction is tangent to the surface, and near such points, the coloring may be inconclusive. For rays in a large open set however, the coloring clearly indicates the presence of rays intersecting the surface a specific number of times.

An example is given by the surface of a revolution of a square. If the axis around which the polygon is rotated is parallel to a side of the square, then the resulting figure is a subset of the union of two closed circular cylinders capped off by circular discs. It follows that the torus thus obtained has geometric order four.

If however we rotate the square forty-five degrees, we obtain a surface of revolution which has geometric order six, a fact which is clearly indicated by the presence of large areas of violet when the projection direction is close to horizontal. Color Plate 18 shows color-by-order images for different projection directions on a torus.

Images from the implicit function renderer were produced at the Brown University computer graphics laboratory. The author would like to acknowledge the continued cooperation of the graphics group under the direction of Andries van Dam, particularly David Laidlaw and David Margolis.

REFERENCES

1. Banchoff, T. and Kuiper, N., *Geometrical class and degree for surfaces in three space*, J. Diff. Geom. **16** (1981), 559–576.
2. Fisher, G. (Editor), "Mathematical Models; Commentary," Vieweg–Verlag, Braunschweig/Wiesbaden, 1986.
3. Goursat, E., *Étude des surfaces qui admettent tous les plans de symétrie d'un polyèdre régulier*, Ann. Sci. Ec. Norm. Sup. **(3) 4** (1987), 159–200.
4. Hirzebruch, F., *Singularities of algebraic surfaces and characteristic numbers*, Max Planck Inst. für Math. (1984 Bonn).
5. Juel, C., *Die elementare ringfläche vierter ordnung*, D. Kgl. Danske Vidensk. Selsk. Skrifter, naturvidensk. og mathem. afd. **8,** Raekke 1.4 (1916), 182–197.

Department of Mathematics, Brown University, Providence, RI 44106

A Low Cost Animation System Applied to Ray Tracing in Liquid Crystals

J. R. BELLARE†, J. A. MACDONALD‡, P. P. BERGSTROM, L. E. SCRIVEN, AND H. T. DAVIS

Abstract. Animated movies of scientific graphics can be recorded on film with the low-cost system of hardware and software described here. The hardware consists of a 16mm camera, a stepper motor, and a simple camera-motor controller. The software is designed to produce bitmaps from graphical data, combine bitmaps into composite frames, and record frames onto film. The camera is fully controlled by the same graphics workstation that is used to display the images, so that fades and dissolves can be done by software with a camera not designed for such special effects. The graphical data, generated on a supercomputer, is subsequently transferred to the workstation where it is stored and recorded frame-by-frame according to a configuration file. A variant of the software, which operates across local and wide area networks, makes use of network computing software to send computationally intensive tasks to a remote supercomputer or to other workstations in a distributed computing environment. We have used the system to simulate polarized light microscope images of liquid crystals according to a single-scattering ray-tracing theory.

§1. Introduction.

Scientific research today often involves simulation of time dependent phenomena wherein the dynamical equations underlying the physical process are solved by supercomputers. Examples include studies of fluid flow in porous media, molecular dynamics of rigid rods, diffusion and aggregation of colloids, and chemical reaction dynamics. The results of such simulations are often best presented as a sequence of graphical output cast in the form of an animated movie.

Equally amenable to animation are results of simulating a process where it is not time but some other parameter that is progressing. The parameter may be a rotation angle, a reaction extent, a volume fraction or mean curvature. In this paper the simulation of interest is image formation in polarized

---

†Present address: Departments of Polymer Science and Engineering, and of Mathematics and Statistics, University of Massachusetts, Amherst, MA 01003

‡Also affiliated with the Institute for Mathematics and Its Applications, Minneapolis, MN 55455

light microscopy of liquid crystals. Animation of the images, as objects rotate and deform, is valuable as a guide to image interpretation. We first review the theory of image formation in polarized optical microscopy and then describe the animation system we developed to present the results as a movie.

§2. Ray Tracing in Liquid Crystals.

A microscope image is the intensity distribution on a viewing plane of a light beam transmitted through the specimen. Thus, simulation of microscope images is equivalent to ray tracing through the specimen. In polarized light microscopy a plane polarized beam of light passes through the specimen and then through an analyzer, after which it forms an image on a viewing screen. Polarized light microscopy is widely used to study the microstructure of crystals, liquid crystals and other optically anisotropic specimens. Ray tracing through ordinary crystals is straight forward because Maxwell's equations describing the transmission of light through such media can be solved with relative ease. Ray tracing through liquid crystals is not easy because their symmetries are complex and the interaction of light with their varied shapes results in complex equations and boundary conditions.

We have recently shown [**1**] that with a single scattering approximation the intensity distribution $\psi(x, y)$ of a light beam of unit incident intensity that has passed through a polarizer with polarization direction $\mathbf{e}_p$, through a specimen of thickness $(z_{far} - z_{near})$ and of local electric susceptibility $\mathbf{X}$, and through an analyzer with analyzer direction $\mathbf{e}_a$ is

$$\psi(x, y) = \left| \int_{z_{near}}^{z_{far}} (\mathbf{I} + C_2\mathbf{X}) : \mathbf{e}_p\mathbf{e}_a dz \right|^2 \tag{1}$$

where $\mathbf{I}$ is the unit dyadic and $\mathbf{X}$, the electric susceptibility, is a dyadic whose eigenvalues $X_i = (n_i^2 - 1), i = 1, 2, 3$ depend on the three principal refractive indices n_i of the specimen.

The equilibrium microstructure of lamellar liquid crystals consisting of elongated molecules is the structure with the lowest free energy. The free energy is lowest when the lamellae are a parallel family of surfaces with zero mean curvature or equivalently, a family of parallel planes of infinite extent. Although there are numerous zero mean curvature surfaces that are multiply connected and curved [**2**], they do not form parallel families. Lamellae of finite size have edges, i.e., crystalline boundaries, which have higher free energy, and so such materials must have closed forms without edges if the edge energy is to be minimized. The competition between zero mean curvature and zero edge energy leads to a compromise between planes and closed forms. Local regions of a material with large mean curvatures are called defects. Defects are not favored because they contribute relatively large free energy per unit volume of the material. Maxwell [**3**] showed that there is only one family of parallel surfaces without surface defects. This is the family of Dupin cyclides. Such a family has two line defects: one is an ellipse or parabola, the other a confocal hyperbola or parabola. A surface of a Dupin cyclide is the envelope of a variable sphere tangent to three fixed spheres [**4**].

Our goal was to calculate images from liquid crystals in the shape of Dupin cyclides, spheroids and ellipsoids, and to illustrate how the images change as the objects rotate in space relative to the polarizer and analyzer. The images change when the objects rotate because the orientation of the electric susceptibility, i.e., its eigenvectors, rotate with respect to the laboratory viewing frame in which Eq. (1) is defined. Thus our simulation evaluated Eq. (1) for the various shapes as they were rotated. Details of the simulation technique are given in [**1**]. The simulation results were gray-scale graphical images that were animated into a movie with the system described next.

§3. The Animation System.

Although there has been much progress in video based animation systems for recording scientific graphics [**5,6**], we chose film as a recording medium. The chief deterrent to film-based animation has been the high cost of commercial film recorders. Hence we elected to build a low-cost system based on a standard 16mm camera coupled to a simple controller. The idea was to aim a tripod-mounted camera at a graphics workstation screen, display a bitmap to be animated and shoot one or more frames of film. The movie was made by shooting a sequence of bitmaps. The camera and computer screen were enclosed in a light-proof chamber assembled from an adjustable frame covered by dark material. The system was built from standard components for $1500, including a camera, controller, light-proof chamber, and tripod. The only additional hardware required was the personal computer or workstation on which the graphics were displayed.

§3.1. Why Film?

Even with the current electronic imaging revolution, traditional silver halide based film remains superior to video as a recording medium. Whereas video has advantages of instant gratification and reusability of media, film beats video on four counts: resolution, dynamic range, single-frame indexing, and longevity. Resolution is the ability to distinguish fine detail. Video tape in the popular NTSC format is limited in resolution to about 300 lines, whereas ordinary 16mm film can resolve more than 1200 lines [**7**,**8**]. There are graphic-arts films that can resolve as many as 3000 lines [**8**]. Most graphics workstations have 1024^2 pixels, so they can display more detailed images than video can record. Dynamic range is the ratio of maximum to minimum intensities that can be faithfully recorded. A large dynamic range is desirable because it is possible to record tonal variations both at low and high intensities. Video is limited to a dynamic range of 2^3 whereas film has a dynamic range of 2^7 [**8**]. A graphics workstation with 24 bits per pixel (eight bits per primary color) has a dynamic range of 2^8. While there are better video formats in vogue (and newer ones are being planned), none yet exceed the resolution or dynamic range of film. Furthermore, the equipment for the better video formats is priced beyond the budget of many researchers. Because standard video recording equipment has no means of indexing to a single frame, special expensive equipment is required for frame-by-frame animation. Film is invariably sprocketed; so it is routinely exposed frame-by-frame in every camera. Videotape is not an archival recording material because its magnetic fields fade with time. When properly processed, film is archival: it is generally expected to last over a hundred years without special care.

Once recorded on film, a movie is easily transferred to video if required for distribution. This indirect route to video—a standard technique in commercial studios for quality TV programming—is superior to direct recording on tape because film compresses the dynamic range of the original graphics so that subsequent transfer to tape does not as greatly exceed its dynamic range. Moreover, 16mm cameras are easily available for a fraction of the cost of a video tape recorder with a single frame controller. New and used

16mm cameras are reviewed and advertised for sale in the publications *American Cinematographer* and *Shutterbug*.

§3.2. Animation Hardware.

The animation system consists of a 16mm movie camera, a stepper motor, and a camera-motor controller. We used a Bolex H16 camera. This camera is designed to be driven by an internal spring, but it also has a shaft extending to the exterior that can be driven by a motor. The shaft not only advances the film but it also opens and closes the shutter. As the shaft rotates at constant angular velocity, the shutter opens, stays open for a certain time interval, then closes, and the film advances. One clockwise turn (360°) of the shaft advances and exposes eight frames. If the shaft is rotated in the opposite direction, the film is exposed as it is rewound, and so if double exposure is desired, it can easily be accomplished.

The shaft is connected to a stepper motor (Vexta model PH265-02B) which rotates by 360° when 200 state transitions are applied to its windings in a particular order known as the drive sequence. Thus 25 state transitions expose and advance one frame of the film. If the drive sequence is reversed, the motor rotates in the opposite direction. The motor is driven by a camera-motor controller (CMC), which in turn is driven by the graphics workstation. The film is advanced or rewound by toggling the state of Data Terminal Ready (DTR, pin 20) on the RS-232C serial port of the workstation. The camera-motor controller is edge triggered. A transition from off to on moves the camera motor one step, and a transition from on to off also moves the motor one step. Whether the film advances or rewinds depends on the logic state of Data Set Ready (DSR, pin 6). If DSR is off, the film advances; if on, it rewinds. To summarize, DTR from the workstation is connected to the camera-motor controller, which converts each DTR transition into a drive sequence to rotate the stepper motor connected to the camera.

To expose one frame, eight DTR transitions, 7ms apart, are output to the CMC. This moves the camera drive train enough to open the camera shutter from a starting position, but not enough to advance the film. The film is exposed for the desired amount of time by suppressing DTR transitions for that period. Then 17 DTR transitions are output. This closes the shutter and advances the film to the next frame. Thus, the speed of film movement, the direction of film travel, and the exposure of the film are under direct control of the workstation. This permits animation software to fade scenes in and out, to dissolve between scenes, to animate periodic events from one period, and to effect slow motion, time lapse and time reversal.

§3.3. Animation Software.

The animation hardware can be used with any workstation or computer that can control signals on an output port via software, because all the CMC needs is at least one signal that can be switched on and off. The signal can come from a RS-232C serial port as described above, or it can come from a parallel port or from a bidirectional port. We have even used a Commodore 64 computer successfully to make animations by controlling a pin on its user port. However, our principal animation software was developed for the Apollo family of graphics workstations. The computer animation system software has three main programs:

1. ***Make_Bitmap,*** converts a numerical-data representation of an image into a bitmap compatible with Apollo Computer's DN580 or DN590 color workstations. It opens a data file that describes an image, creates a bitmap, and then saves the bitmap to disk. Currently two different data-file formats are supported. The first format represents a two-dimensional array of gray-scale data. The second format is an SRL file produced by Movie.byu [**9**]. Make_Bitmap takes as input a configuration file containing data-file types, data-file names, and bitmap-file names.
2. ***Make_Frame,*** creates the image to be captured on film by merging text and bitmaps. It takes the bitmaps saved by Make_bitmap and places them on the screen at coordinates supplied by the user. The user can then interactively add text anywhere in the frame buffer by using the mouse and keyboard. At the end of the program the screen is saved on the disk drives as a 1280 by 1024 bitmap file. Make_Frame takes as input a configuration file which defines the bitmap-file names and locations, and several constants used to fill the color map.
3. ***Umn.Movie,*** uses the bitmaps created by Make_Frame or Make_Bitmap to make a movie scene by repeatedly placing bitmaps on the computer screen and signaling the movie camera to advance one or more

frames. Currently this program optionally fades in on a display, shoots a series film frames, then optionally fades out. As do the other programs, this one reads configuration data from an input file. Umn.Movie puts a picture on the screen, opens the camera shutter by cycling DTR on the serial port, leaves the shutter open for the exposure length of time, and then closes the shutter and advances the film by cycling DTR.

An animated movie is made by generating a series of frames and calling Umn.Movie to film the series in sequence. Each frame can consist of one or more bitmaps. A shell script can be used to call Umn.Movie with different configuration files to make several movie sequences ("scenes"). Alternatively, several scenes can be specified in one configuration file with fades and dissolves. Thus, one call to Umn.Movie can film the entire movie specified in a configuration file without any need for further human intervention. More details about the software and hardware, including circuit diagrams and printed circuit layouts, are available [**10**].

§3.4. Network Computing.

The animation system described above has one major limitation. Each bitmap is a 1280-by-1024 array of 24-bit integers (3.75 Mb. of data). So one second of film (at 24 frames per second and 3 frames per bitmap) can require as much as 30 Mb. of disk storage. This is a worst-case example and storage requirements can be reduced by not saving the background data or by using various data compression techniques [**6**]. To have a system which makes a movie of arbitrary length without human intervention would require an unreasonably large amount of disk storage. A solution to this problem was found by using Apollo Computer's Network Computing System (NCS) [**11**]. NCS is made up of two components: the network interface definition language (NIDL) compiler, and the network computing kernel (NCK). An interface definition is written in NIDL syntax and passed through the NIDL compiler to create several include files and stub procedures for distributed applications. Stub procedures are the routines that make remote procedure calls look almost local. With NIDL, programmers do not have to handle details like data format conversions across a heterogeneous network of computers. NCK is the run-time library that finds resources on the network by using a location broker.

Without NCS, the computationally intensive procedures (which create the numerical-data representations of an image) are done on a supercomputer and saved on magnetic tape. The tapes are then restored on the Apollo disk drives until available free space is exhausted. That portion of the data is run through Make_Bitmap, Make_Frame, and Umn.Movie and thus recorded on film. The old bitmaps and data files are then deleted and replaced with data for the next run (this step requires human intervention).

With NCS a remote procedure call is made to the supercomputer to generate the data needed for one workstation display. Those data are then processed by Make_Bitmap, Make_Frame, and Umn.Movie and the whole process repeated for the next display. In this way very few data files and bitmaps need to be saved (and can be saved as temporary files).

§4. Results and Discussion.

Figure 1 shows one frame of the animated movie, which was recorded on Kodak 7292 with a 25mm f/1.4 Switar lens exposed for 2 sec at f/16 per frame and 3 frames per computer image. The Apollo version of the software and the movie (recorded in color on 16mm film and 3/4″ U-matic (NTSC) or 1/2″ VHS (NTSC) videotape format), are available for loan from the authors. The figure shows polarized microscope images from a variety of objects selected to illustrate different possible shapes of liquid crystals: oblate spheroids, prolate spheroids, ellipsoids, fused spheres of equal diameters, fused spheres of unequal diameters, ring torus, spindle torus, and three types of Dupin cyclides. The movie clearly depicts the changes in the images as the objects are rotated, thereby providing a guide for interpreting images from polarized light microscopy.

The animation system permits recording of graphics at high resolution on film with equipment that costs about an order of magnitude less than equivalent video equipment. The system is so flexible that it can be used with any computer that can toggle an output pin. Because the speed and direction of film transport can be controlled by the software, special effects can be generated easily. The success of the system, we believe, is clearly demonstrated in the movie we made. When used in a distributed computing environment with network computing software, it becomes a powerful yet affordable tool to produce animations of supercomputer simulations efficiently.

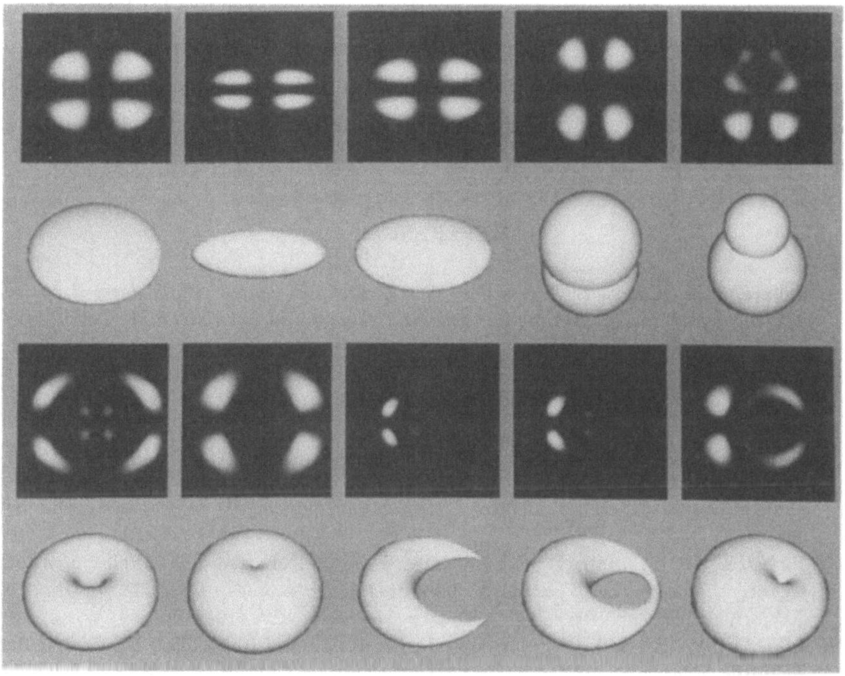

Figure 1: One frame of the movie generated using the animation system. It depicts liquid crystals of different shapes and the results of a special case of ray tracing through them. The special case is that of orthogonal polarizer and analyzer, which makes the ray tracing results simulations of cross-polarized optical microscope images of the objects. The objects are shown in perspective view along the same viewing direction as the images depicted below the object. The objects are (in left-to-right and top-to-bottom order) oblate spheroid, prolate spheroid, ellipsoid, fused spheres of equal diameters, fused spheres of unequal diameters, ring torus, spindle torus, horned Dupin cyclide, ring Dupin cyclide, and spindle Dupin cyclide.

Acknowledgements.

Financial support for this work came from the United States Department of Energy (Grant DOE/DEAC 19-79BC10116-A013) and the Minnesota Supercomputer Institute. We thank A. R. Anderson and W. R. King of the Minnesota Supercomputer Center for assisting in optimizing the simulation program, T. Heisler of Apollo Computer Inc. for the loan of a DN590 workstation, and Mr. and Mrs. Donald Swadner for construction of the light-proof chamber.

REFERENCES

1. Bellare, J. R., Davis, H. T., Miller, W. G., and Scriven, L. E., *Polarized optical microscopy of anisotropic media: Imaging theory and simulation*, J. Colloid and Interf. Sci. (1990), (in press).
2. Anderson, D. M., Davis, H. T., Nitsche, J. C. C., and Scriven, L. E., *Periodic surfaces of prescribed mean curvature*, Advances in Chemical Physics (1989), (in press).
3. Maxwell, J. C., *On the cyclide*, Quart. J. Pure and Appl. Math. 9 (1869), 111–126.
4. Dupin, Charles, *Applications de geometrie et de mechanique, a la marine, aux ponts et chaussees, etc., pour faire suite aux developpements de geometrie*, Batchelier, Paris (1822).
5. Johnston, W. E., Hall, D. E., Renema, F., and Robertson, D., *Principles and techniques for low cost computer generated video movies, LBL-22330*, Univ. of California, Lawrence Berkeley Laboratory, Berkeley, CA (1987).
6. Johnston, W. E., Hall, D. E., Huang, J., Rible, M., and Robertson, D., *Distributed scientific video movie making, LBL-24996*, Univ. of California, Lawrence Berkeley Laboratory, Berkeley, CA (1988).
7. Neblette, C. B., "Photography: Its Materials and Processes," 5th ed. Van Nostrand (New York), 1952, p. 317.
8. Higgins, G. C., *Information capacity of photographic materials*, in "Photographic Systems for Engineers," ed. by Brown, F. M., Hall, H. J., and Kosar, J., Society of Photographic Scientists and Engineers (Washington D.C.) (1966), p. 202.
9. Christiansen, H., Stephenson, M., Nay, B., and Grimsrud, A., "Movie.byu Training Text," 2nd edition, Bookcrafters, 1987.
10. MacDonald, J. A., Bergstrom, P. P., Bellare, J. R., Scriven, L. E., and Davis, H. T., *A low-cost film-based animation system for visualizing scientific data utilizing distributed computing*, (in preparation).
11. Apollo Computer Inc., "Network Computing System Reference," Order No. 010200 Revision 00, Chelmsford, MA 01824, 1987.

Department of Chemical Engineering and Materials Science, University of Minnesota, Minneapolis, MN 55455

From Sketches to Equations to Pictures: Minimal Surfaces and Computer Graphics

MICHAEL J. CALLAHAN

In 1984, mathematician David Hoffman and computer scientist Jim Hoffman (no relation) sat before a computer screen and saw a crude rendition of a new minimal surface unfold itself. Inspired by the computer graphics, David Hoffman and William Meeks III were able to prove that this surface, which had been discovered by C. Costa, was embedded [**5**][**7**]. The surface then joined the classical examples of the plane, the catenoid and the helicoid as the only known examples of complete, embedded minimal surfaces of finite topological type.

The story of this initial discovery has been told elsewhere [**6**], and the proof of embeddedness appears in [**7**]. In the years since this discovery, both the mathematical tools for constructing complete, embedded minimal surfaces and the computer tools for visualizing them have improved. The purpose of this paper is to describe some of those mathematical and computational tools. We first sketch a simple technique for writing down analytic representations of possible complete, embedded minimal surfaces; this technique easily motivates the equations of all the examples of complete embedded minimal surfaces of finite type which have been discovered since Costa's surface. We then describe briefly the computer programs which have been developed to allow these equations to be turned easily into coherent computer graphics.

The Analytic Representation.

The original surface of Costa and the examples discovered subsequently are presented using the analytic Enneper–Weierstrass representation. This representation describes a minimal surface in $\mathbf{R}^3$ as the image of a conformal, minimal immersion of a Riemann surface and, conversely, gives a procedure for constructing such immersions of Riemann surfaces. Suppose M is a Riemann surface. Let g be meromorphic function on M, and η be a

The research described in this paper was supported in part by the Department of Energy (DE-FG02-86ER25015) and the National Science Foundation (DMS-8503350).

holomorphic 1-form with zeros precisely at the poles of g, at which points the zeros have twice the order of g's poles. Then the map $X: M \to \mathbf{R}^3$ defined by

$$X(z) = \mathrm{Re} \int_{z_0}^{z} \begin{matrix} (1-g^2)\eta \\ i(1+g^2)\eta \\ 2g\eta \end{matrix}$$

(where $z, z_0 \in M$, z_0 fixed) is a conformal minimal immersion, which is well-defined provided that the periods of the three 1-forms appearing in the integral are purely imaginary. Furthermore, the meromorphic function g, viewed as a map to the Riemann sphere $\mathbf{C} \cup \{\infty\}$, is the Gauss map of the surface; to be precise, if $G: M \to S^2$ is the usual Gauss map obtained by translating the normal vectors to the surface to the origin of $\mathbf{R}^3$, and if $\sigma: S^2 \to \mathbf{C} \cup \{\infty\}$ is stereographic projection, then $g = \sigma \circ G$. (See **[12]** for more information on the Enneper–Weierstrass representation.)

The technique for writing down equations for a complete embedded minimal surface consists of deducing its Enneper–Weierstrass representation from rough geometric data like its symmetry group and its asymptotic behavior. In order to proceed, we need to make the additional assumption that M has finite total curvature.[1] Before describing the procedure in more detail, we need some additional information about the behavior of M, g and η for complete embedded minimal surfaces of finite total curvature.

Osserman **[11]** proved that if M is a complete minimal surface of finite total curvature, then: M is conformally diffeomorphic to a compact Riemann surface $\bar{M}$ punctured at a finite set of points $p_1, \ldots, p_n$; g and η extend meromorphically to $\bar{M}$; and the immersion X of M into $\mathbf{R}^3$ is proper. Thus in our case, we may, in writing down g and η, work on the compact Riemann surface $\bar{M}$, and the punctured neighborhoods of the $\{p_i\}$ will then correspond to the ends of the surface. We can thus refer loosely to "the end p_i", meaning the end of M having a punctured neighborhood of p_i as a representative.

The geometric behavior of the end p_i is determined by the analytic behavior of g and η at the point p_i. We refer to the value of g at a point p_i as the *limiting normal vector* at the end p_i. It is easy to prove that the limiting

[1] Here we do not distinguish between M and its image $X(M)$; we say that M has finite total curvature if $\int_{X(M)} K \, dA$ is finite, where K is the Gaussian curvature. The assumption that M has finite total curvature implies that M has finite topology; it is unknown whether there exists a complete embedded minimal surface of finite topology and infinite total curvature other than the helicoid. From here on, we will identify M with its image.

normals of a complete embedded minimal surface of finite total curvature are parallel [**10**]. Hence we may, after a possible rotation in $\mathbf{R}^3$, take g to have values 0 and ∞, corresponding to vertical limiting normal vectors $(0, 0, \pm 1)$, at the points $\{p_i\}$. Further, every end of a complete embedded minimal surface of finite total curvature is asymptotic either to a catenoid or to a plane (see, e.g., [**10**], [**13**]). The end p_i is asymptotic to a catenoid if g has a simple pole or zero at p_i; otherwise, it is asymptotic to a plane. Finally, the order of η at a point p_i is forced by the requirement that the first two forms in the Enneper–Weierstrass equation have poles of order 2 at an end; this requirement follows from embeddedness and completeness. (Note that η, while holomorphic on M, may have poles at the $\{p_i\}$.)

Notice that we now have a rough geometric description of the pole structures of g and η. On $M \subset \bar{M}$, g has zeros and poles wherever the Gauss map is vertical, and η is holomorphic with zeros precisely at g's poles, with twice their order. At the points $\{p_i\} = \bar{M} - M$, g must have a zero or a pole, and g is simple at ends asymptotic to catenoid ends and has higher order otherwise. The order of η at each p_i is determined by the order of g there.

We can now write down candidates for complete embedded minimal surfaces of finite total curvature. We imagine the surface, specifying its topology, its symmetries, its end behavior, and its vertical normals. We can then produce the data for its Enneper–Weierstrass representation. From its topology we write down the Riemann surface M; we can then apply the rules above to deduce the divisors of g and η on $\bar{M}$ from the geometric information about the surface's ends and vertical normals. Generally, there will be parameters in the equations which specify, for example, the conformal structure of M, the location of vertical normals on M, and constants in g and η. The question of existence of the hypothesized example is then reduced to the problem of proving that choices for these parameters exist such that the periods of the forms in the Enneper–Weierstrass equation are purely imaginary. (Of course, if such choices exist, it remains to prove that the resulting complete immersion $X: M \to \mathbf{R}^3$ is, in fact, an embedding.)

There is one more assumption we must make if this technique is to be feasible in practice. The problem is that the procedure, as described, is too general: for example, the distinguished points on $\bar{M}$ where g and η have their poles and zeros—that is, the points in M with vertical normals and the ends $\{p_i\}$—could, in general, be located anywhere on $\bar{M}$, although sometimes the surface's symmetries can help determine where these points lie. Furthermore, it can be difficult in the first place to write down the Riemann surface $\bar{M}$.

To solve these problems, we make the additional assumption that M is symmetric under a symmetry group $\mathbf{R}$ of rotations around a vertical axis in $\mathbf{R}^3$ (that is, an axis parallel to the limiting normal vectors of the ends of M). This assumption holds for all known examples of complete embedded minimal surfaces of finite total curvature. The surface M is a branched cover of $M/\mathbf{R}$. This covering may be extended to a branched cover $\bar{M} \to \bar{M}/\mathbf{R}$ since $\mathbf{R}$ fixes each end of M. Let $\bar{N} = \bar{M}/\mathbf{R}$. The advantage of making this assumption is that the topology of $\bar{N}$ is often simpler than that of $\bar{M}$; in fact, for the known examples, $\bar{N}$ is simply S^2. In addition to simplifying topological issues, introducing the symmetry group $\mathbf{R}$ provides information about the pole structures of g and η. The branch points of the covering consist of the ends $\{p_i\}$ and the points in M on the axis of the rotation group $\mathbf{R}$. As observed above, the orders of g and η at the points $\{p_i\}$ are determined by the geometric behavior of the corresponding ends; the presence of the rotational symmetries $\mathbf{R}$ constrains these orders. The normal vectors at the points where M meets the axis must be vertical, if M is to be embedded; hence g has poles or zeros there too, and the order of g is again constrained by the rotational symmetries.

We would like to be able to work on $\bar{N}$ rather than $\bar{M}$, because of its simpler topology. To do this, we must be able to recover $\bar{M}$ from $\bar{N}$. We can do this using the Gauss map g. Note that the values of g on a fiber over a point in $\bar{N}$ consist of an orbit of the induced action of $\mathbf{R}$ on the Riemann sphere of normal vectors. Since this induced action is generated by multiplication by $e^{2\pi i/k}$, where k is the order of the rotation group $\mathbf{R}$, the function g^k is well-defined on $\bar{N}$. Of course, the function g^k on $\bar{N}$ can be determined by the geometric data describing the behavior of ends and the vertical normals on the surface. Furthermore, the covering produced by taking the k-th root of the function g^k on $\bar{N}$ is equivalent to the covering $\bar{M} \to \bar{N}$. Thus given $\bar{N}$ and g^k we can recover $\bar{M}$.

Our technique is as follows: We first suppose that the surface is symmetric under a rotation group $\mathbf{R}$ of order k around a vertical axis. Then, using information about the appearance of vertical normals on the surface and behavior of ends, we write down an equation for the function g^k on the quotient surface $\bar{N}$, whose topology is assumed to be known; the symmetries of the surface are helpful here in placing the poles and zeros of g^k. Finding g^k on $\bar{N}$ immediately provides the Riemann surface $\bar{M}$ and the Gauss map g on $\bar{M}$. Finally, we write down the meromorphic 1-form η on $\bar{M}$. This completes the Enneper–Weierstrass representation for the candidate surface; existence again requires that real parts of periods be killed off. This technique is described in detail in [**2**].

As an example, we here apply the procedure to the Hoffman–Meeks examples M_k, $k \geq 1$ [**8**]. The Costa surface appears as the first surface in this family. The surface M_k has genus k and 3 ends,[2] and this implies that the total curvature of M_k is $-4\pi(k+2)$ [**8**], [**10**]. One of the ends is asymptotic to the plane $\{x_3 = 0\}$ and the other two are asymptotic to catenoid ends. The plane end has limiting normal vector $(0,0,1)$, and the catenoid ends have limiting normal vector $(0,0,-1)$. The surface M_k has a rotational symmetry group $\mathbf{R}$ about the x_3-axis of order $k+1$, and $M_k/\mathbf{R}$ is conformally the Riemann sphere $\mathbf{C} \cup \{\infty\} = S^2$ punctured at three points corresponding to the ends of M_k. The surface M_k contains one point with vertical normal vector; this point is the origin $\vec{0}$ and hence lies on the axis of the rotation group $\mathbf{R}$. The normal vector at $\vec{0}$ is $(0,0,-1)$. The intersection $M_k \cap \{x_3 = 0\}$ consists of $k+1$ equally-spaced lines meeting at the origin, and M_k is symmetric under rotation by π around any of these lines. Finally, M_k is symmetric under reflection through the $k+1$ equally-spaced planes containing the x_3-axis and bisecting the $k+1$ lines in $\{x_3 = 0\}$. See Color Plate 15 for an image of the surface M_2.

Now, let us use this information to derive the Enneper–Weierstrass representation of M_k. By possibly applying a linear-fractional transformation, we may assume that $M_k/\mathbf{R}$ is $\mathbf{C} \cup \{\infty\} = \bar{N}$ punctured at the points $\{0, 1, \infty\}$ and that the point 1 corresponds to the plane end. We will use the symmetries of the surface to show that the point p corresponding to the axial point $\vec{0}$ is -1. First observe that the rotations about the lines in $\{x_3 = 0\}$ induce a unique action on $\bar{N}$, consisting of an anticonformal diffeomorphism fixing 1 and interchanging 0 and ∞; this clearly consists of inversion through the unit circle. Since the rotations also fix $\vec{0}$, we must have $|p| = 1$. The reflections through vertical planes similarly induce a single action on $\bar{N}$. This action is orientation-reversing and fixes 0, 1 and ∞; it is thus complex conjugation. Hence $\operatorname{Im} p = 0$; since 1 is the plane

[2] That is, M_k is conformally diffeomorphic to a compact surface of genus k with three points removed.

end, the only possibility is $p = -1$.

The branch points of $\bar{M}_k \to \bar{N}$ are thus $\{0, \pm 1, \infty\}$. Since M_k had only one point with vertical normal, $\vec{0}$, we can now write down g^{k+1} on $\bar{N}$. Here we use rules concerning the behavior of g^{k+1} on the quotient surface which are described in detail in [**2**]. At the catenoid ends 0 and ∞, g^{k+1} must have simple zeros. At the plane end 1, g^{k+1} must have a pole of order $m(k+1)+1$, where $m \geq 1$; since, by the total curvature of M, g has degree $k+2$, $m = 1$. At the axial point -1, g^{k+1} must have a zero of order $m(k+1)-1$, $m \geq 1$; again $m = 1$ by degree considerations. Hence we have an equation for g^{k+1} on $\bar{N}$:

$$g^{k+1} = c\frac{z(z+1)^k}{(z-1)^{k+2}}$$

where c is a complex constant to be determined. This equation defines a Riemann surface $\{(z, g)\} \subset (\mathbf{C} \cup \{\infty\} - \{0, 1, \infty\}) \times (\mathbf{C} \cup \{\infty\})$, which is conformally equivalent to M_k. On this surface we can write down an equation for η, after deducing its pole structure. Since g has no poles on $\mathbf{C} \cup \{\infty\} - \{0, 1, \infty\}$, η is regular on the lift of that set to M_k. At the lifts of 0 and ∞, η must have double poles. At the lift of 1, η must have a zero of order equal to $2k+2$. (Again we are using rules which are described in more detail in [**2**].) Since the lift of the 1-form dz has zeros of order k at 0 and ± 1 and a pole of order $k+2$ at ∞, we find that

$$\eta = c'\frac{dz}{zg}$$

where c' is a complex constant to be determined. This completes the derivation of the Enneper–Weierstrass representation of the surface M_k.

Similar analysis can be applied to find the Enneper–Weierstrass representations of the other known complete embedded minimal surfaces of finite total curvature: the smooth perturbations of M_k found by Hoffman and Meeks [**9**] and the four-ended surfaces found by Callahan, Hoffman and Meeks [**4**]. In addition, Callahan, Hoffman and Meeks [**2**] used essentially this method to derive the analytic representation of a family of surfaces $\tilde{M}_k$ having a single translational symmetry and an infinite number of ends; these examples join a continuous family of surfaces found by Riemann as the only known examples of singly-periodic complete embedded minimal surfaces having an infinite number of ends. Color Plate 16 depicts the surface $\tilde{M}_1$.

Solving for the Surface.

The development of analytic techniques for constructing embedded minimal surfaces has proceeded hand-in-hand with the refinement of the computational tools available for their study. When D. Hoffman, J. Hoffman and W. Meeks originally studied the example discovered by Costa, it was presented as the immersion of a square torus $\mathbf{C}/\{m + ni \mid m, n \in \mathbf{Z}\}$ punctured at the points $\{0, 1/2, i/2\}$, and g and η were given in terms of the Weierstrass $\wp$-function and its derivative. The method used to compute a discrete representation of the surface for graphing was simple: a program computed the values of the map X at points lying on a rectangular grid contained in $\{z \mid 0 < \operatorname{Re} z < 1, 0 < \operatorname{Im} z < 1\}$, and these values were then fed to a graphics program which drew the rectangles they defined.

This method has two significant problems. First, it ignores the symmetries of the surface, and therefore wastes computation. Second, it ignores the metric of the surface, and the grid of points it produces therefore become widely spaced out in some regions and too closely spaced in others; indeed, the first pictures produced were quite hard to interpret for precisely this reason. When the Costa surface was first studied, these problems were practically unavoidable, since the symmetries of the surface were unknown and the computational tools available were crude. For the surfaces discovered since then, these two problems are more acute, and this made finding a method of producing the discrete graphical representation of the surfaces crucial. In the subsequent examples, the computational waste involved in computing the whole surface is even greater, due to the greater amount of symmetry. And even if this wasted computation were acceptable, it would still be difficult to write down a parameterization of the entire surface for examples with more complicated topology.

These problems have been solved by a program called `MESH` which automatically generates triangulations. The triangulations `MESH` generates can then be used by the graphics program to produce drawings of the surface. `MESH` triangulates the lift of a planar region onto the surface, choosing the vertices of the triangulation so that its edges have roughly constant length. That is, `MESH` places vertices close together in the parameterization domain where the metric gets large, so that the edges in $\mathbf{R}^3$ are all of reasonable length. (In fact, the program can be tuned so that it places points more closely together where the surface is highly curved, producing a better approximation to the surface.) This scheme works well for the examples mentioned in this paper. For each of these, the quotient surface $M/\mathbf{R}$ of M modulo rotations around the vertical axis is conformally equivalent to the Riemann sphere punctured at a finite number of points. In addition, each of these surfaces has reflectional symmetries through vertical planes which act on $M/\mathbf{R}$ by complex conjugation. Thus the lift of, for instance, the upper half-plane is a symmetric piece of the surface and is a natural choice for the parameterization region. Figure 1 shows a basic piece of a 4-ended complete embedded minimal surface as produced by MESH. (On the computer screen MESH displays lines in different colors, depending on details of how they are generated. Because of this, some of the lines appear faint in this black-and-white image.)

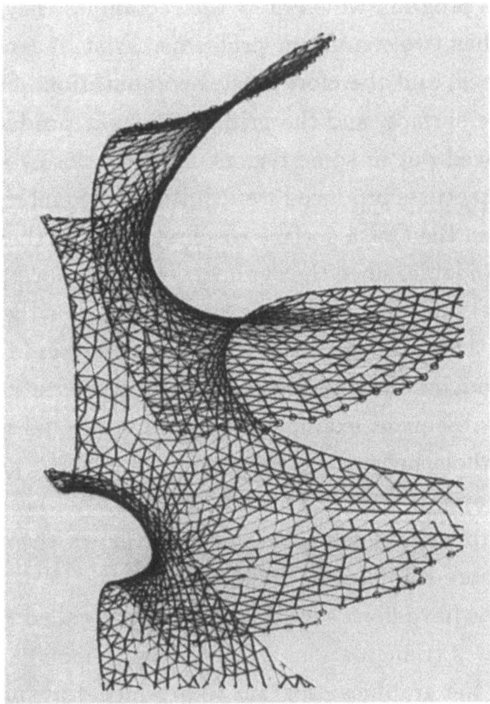

Figure 1

The implementation of MESH has two parts. The triangulation itself is performed by the main program, designed and implemented by James Hoffman. This part is written in C and contains the algorithms for placing points, for drawing crude pictures of the triangulation in the parameterization domain and in space (useful for tuning the parameters of MESH which control how fine a triangulation to produce), and for writing the triangulation to disk in a format suitable for input to the graphics program. However, this part contains none of the code for actually computing points on the surface in $\mathbf{R}^3$.

That is done by the second part of `MESH`, consisting of a set of functions written in `FORTRAN` designed and implemented by the author. Each family of surfaces is presented to the `C` part of `MESH` as two `FORTRAN` functions. One, the setup function, is called once at the beginning of a run. The `C` code passes to this function information obtained from the user about which surface to compute. For example, for the examples M_k described above, this consists of the integer k. The setup function performs calculations which must be done before points of the surface can be computed. Usually this involves finding the values of parameters in the Enneper–Weierstrass equation which kill off real periods, either by computing these values on-the-fly in simple cases like the surfaces M_k or by looking them up in a file of pre-computed values. The setup function then passes back to the `C` code information describing the parameterization domain and a point at which to start the triangulation.

The program then lays down the triangulation by progressively working outward from the start point. To do this it uses the second `FORTRAN` function, which given two nearby points in the parameter space performs a path integral from one to the other and returns the difference vector between the two corresponding points in $\mathbf{R}^3$. As the triangulation is built up, the `C` code uses this function to probe points just outside the existing triangulation to find one which can be added without producing any too-long edges. When the code adds a vertex to the triangulation, it stores the position of the corresponding point in space, along with the normal to the surface there (this is used for making sure that edges are smaller where the surface is highly curved and is also useful in later rendering the surface in the graphics program). In addition, the `C` code can store a piece of information at the vertex which it does not attempt to interpret, but which it will pass to the integration function when it integrates from this vertex to another point. Typically this piece of information describes the lift of the vertex onto the Riemann surface. This information is useful when the integration function has to perform a path integral from the vertex to a nearby point; knowing which lift of the vertex was chosen allows a curve between the points to be lifted to the Riemann surface in order to perform the path integral.

MESH allows new surfaces to be tried out quickly and with great confidence that the resulting triangulation will accurately approximate the actual mathematical surface. Usually the amount of code required to implement a new surface family is less than two printed pages; in part, it is this short because it can use subroutines shared with other surface families for performing path integrals, converting values of g to normal vectors, and so on. MESH thus provides a reliable and easy-to-use tool for converting the Enneper–Weierstrass representation of a minimal surface into a discrete triangulation a symmetric piece of the surface in space.

Drawing the Picture.

Yet it remains to turn this triangulation into an interpretable picture. The program for doing this is called VPL. VPL, which stands for Visual Programming Language, is in fact a toolkit created by Jim Hoffman for designing application-specific very-high-level languages. It consists of a core program to which one can add data types and operators which act on these data types; the data types and the operators are designed to make it easy to write useful programs for the given application in the resulting language. For example, the version of VPL used for graphics has as one of its most-used types *set of triangles* or *triangle set*, which is what MESH produces as output. Operators in VPL can rotate, reflect or translate a triangle set, clip a triangle set by a sphere or a plane, merge multiple triangle sets into one, paint colors on each vertex of a triangle set, or render a triangle set on a graphics device.

VPL provides a visual editor which allows inexpert users to construct programs for displaying surfaces. A program in VPL consists of a tree of operators; the data in the program flows from the leaves of the tree toward the root. Usually, one leaf of the tree is an operator that retrieves a triangle set, which was produced by MESH from a file. The triangle set is then reflected and rotated to produce the whole surface, and is finally drawn by a rendering operator at the root of the tree. In the visual editor, a user can construct this program simply by using a mouse to select from menus. The tree is drawn as a set of nested boxes representing operators. Figure 2 shows a typical VPL editor screen. Each box has slots corresponding to the operator's inputs; the operators which produce the inputs are displayed inside the slots. To construct a program, one fills input slots with operators or, for slots which accept certain simple data types, constants. To fill a slot, one simply clicks on it with a mouse. VPL then presents a menu of operators which output the appropriate data type for the slot being filled. In addition, if appropriate, VPL offers the option of inserting a constant. One either selects an operator or enters a constant. Notice that in any case there is no possibility of syntax or type-mismatch errors.

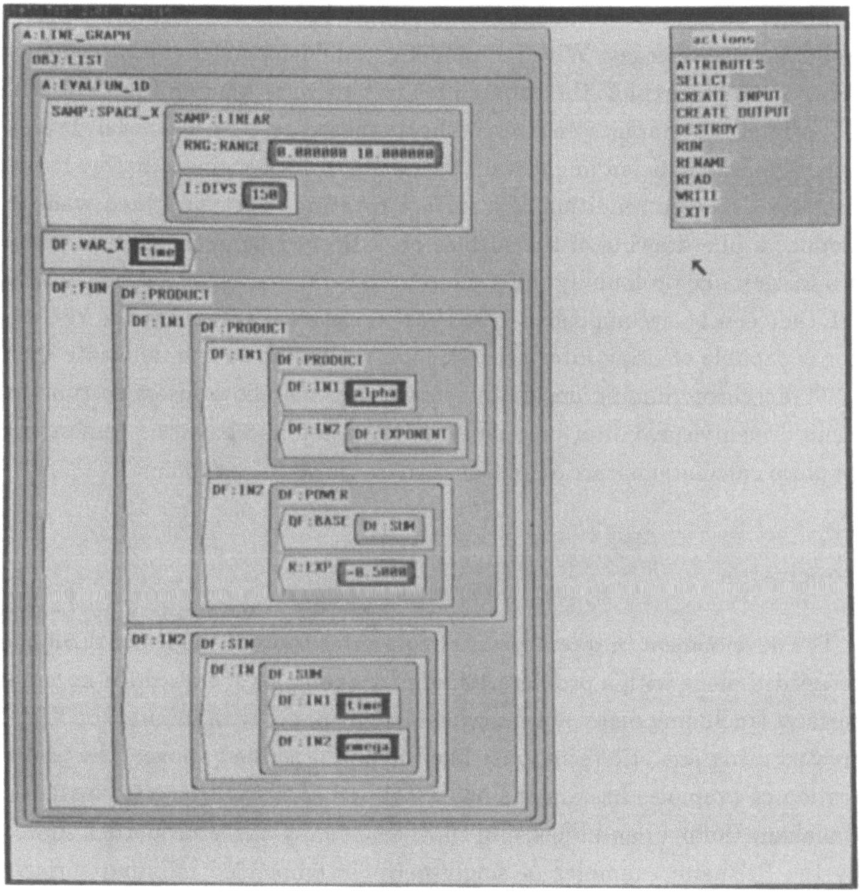

Figure 2

Once the user has constructed a program in the editor, it can be executed simply by selecting a menu option *Run*. VPL programs are interpreted, so no compilation is necessary. (VPL itself is not interpreted, of course; it is written in C.) This allows VPL to be used as an exploratory tool. If the user wants to see the surface, say, from a different angle or clipped by a reflectional plane of symmetry, all that is required is to interpose a rotation or clipping operator in the tree and run the program again. This flexibility comes at relatively little expense. Since a typical operator in a VPL program might rotate 2,500 triangles and the normal vectors at their vertices, the computational overhead of interpretation is minimal.

VPL has, finally, some technical features which make it particularly useful for inexpert users. The range of its rendering operators allows similar VPL programs to produce everything from simple hidden-line drawings to colored ray-traced images. With its sophisticated device-independent graphics library, VPL can exploit the capabilities of hardware ranging from dot matrix printers to graphics engines with z-buffered screens and local display lists. A mathematician might want to use a graphics engine's display list to visualize a colored rendition of a surface rotating in $\mathbf{R}^3$, and then want to produce a line drawing of the surface on a Postscript laser printer. These two tasks make profoundly different demands on graphics software, but in VPL they can be accomplished in similar, simple ways. Finally, the VPL editor is capable of using inter-process communication to communicate with a VPL executor running on another machine; this allows users to run the menu-driven visual editor on a small workstation and have the demanding graphics calculations carried out on a more powerful machine.

Conclusion.

The development in recent years of computer tools like VPL and MESH has proceeded along with a proliferation of new examples of embedded minimal surfaces (including many not described in this paper with infinite topology) created using analytic techniques like the one described above. The proliferation of examples has, on the one hand, led to new theory; for instance, Callahan, Hoffman and Meeks [**3**] show that many of the properties shared by the Riemann examples of singly-periodic embedded minimal surfaces with infinitely many ends and the newly discovered examples are, in fact, forced. Yet perhaps more surprisingly, the ability to create clear graphical pictures has produced new interactions with fields outside mathematics; see, for example, the description in [**1**] of the application of minimal surfaces to polymer physics. These two developments—new theory and new connections—may suggest some of the benefits of using computers to go from equation to pictures.

Acknowledgements.

The graphical images accompanying this paper were produced by the programs VPL and MESH, tools which owe their existence above all to the extraordinary effort and programming expertise of Jim Hoffman. I would also like to thank David Oliver and Jim Riordan for help in preparing this paper.

REFERENCES

1. M. Callahan, D. Hoffman and J. Hoffman, *Computer graphics tools for the study of minimal surfaces*, Communications of the ACM **31** (1988), 648–661.
2. M. Callahan, D. Hoffman and W. Meeks III, *Embedded minimal surfaces with an infinite number of ends*, to appear, Invent. Math..
3. M. Callahan, D. Hoffman and W. Meeks III, *The structure of singly-periodic minimal surfaces*, preprint.
4. M. Callahan, D. Hoffman and W. Meeks III, *Embedded minimal surfaces with four ends*, preprint.
5. C. Costa, *Example of a complete minimal immersion in* $\mathbb{R}^3$ *of genus one and three embedded ends*, Bull. Soc. Bras. Mat. **15** (1984), 47–54.
6. D. Hoffman, *The computer-aided discovery of new embedded minimal surfaces*, Math. Intell. **9**, 3 (July 1987), 8–21.
7. D. Hoffman and W. Meeks III, *A complete embedded minimal surface with genus one, three ends and finite total curvature*, J. Differ. Geom. **21** (1985), 109–127.
8. D. Hoffman and W. Meeks III, *The global theory of embedded minimal surfaces*, preprint.
9. D. Hoffman and W. Meeks III, *One-parameter families of embedded minimal surfaces*, preprint.
10. L. Jorge and W. Meeks III, *The topology of complete minimal surfaces of finite total Gaussian curvature*, Topology **22** (1983), 203–221.
11. R. Osserman, *Global properties of minimal surfaces in* E^3 *and* E^n, Ann. of Math. **80** (1964), 340–364.
12. R. Osserman, "A Survey of Minimal Surfaces," 2nd edition, Dover Publications, New York, 1986.
13. R. Schoen, *Uniqueness, symmetry, and embeddedness of minimal surfaces*, J. Diff. Geom. **18** (1983), 791–809.

Department of Mathematics, Science Center, One Oxford Street, Cambridge, Massachusetts 20138; Internet callahan@zariski.harvard.edu

Nonuniqueness and Uniqueness of Capillary Surfaces

PAUL CONCUS AND ROBERT FINN

ABSTRACT

It is shown that for any gravity field g and contact angle γ, an axially symmetric container can be found that differs arbitrarily little from a circular cylinder and can be half filled with liquid in a continuum of distinct ways, such that no two of the surface interfaces are mutually congruent and such that all of them are in equilibrium with the same mechanical energy. This answers affirmatively a question raised by Gulliver and Hildebrandt [**1**], who obtained such a container in the particular case $g = 0$, $\gamma = \pi/2$.

For a particular surface in the continuum, it is shown that the second variation of energy can be made negative by a non-axisymmetric perturbation under the volume constraint, and thus that the surface can be embedded in a one-parameter family of nonsymmetric surfaces bounding constant volume, with decreasing energy. As a consequence, a rotationally symmetric container deviating arbitrarily little from a circular cylinder is characterized, so that an energy minimizing configuration filling half the container exists but will not be symmetric.

Finally, conditions on the container curvature are given under which the symmetric configurations with prescribed volume are uniquely determined. In the special case for which the container is a sphere and $g = 0$, the symmetric configuration is unique and energy minimizing among all possible configurations.

Details of this work are given in [**2**] and in [**3**].

REFERENCES

1. R. Gulliver and S. Hildebrandt, *Boundary configurations spanning continua of minimal surfaces*, Manuscr. Math. **54** (1986), 323–347.
2. R. Finn, *Nonuniqueness and uniqueness of capillary surfaces*, Manuscr. Math. **61** (1988), 347–372.
3. P. Concus and R. Finn, *Instability of certain capillary surfaces*, Manuscr. Math. **63** (1989), 209–213.

Lawrence Berkeley Laboratory and Department of Mathematics, University of California, Berkeley, CA 94720 and Mathematics Department, Stanford University, Stanford, CA 94305

Static Theory of Nematic Liquid Crystals

J. L. ERICKSEN

ABSTRACT

This will be an expository lecture on liquid cyrstal theory, covering some of the physical phenomena of interest, the formulation of related mathematical problems, and recent mathematical work concerning analysis of them, as well as numerical computations.

More commonly considered static problems involve minimizing energies of the form

$$\int_{\Omega} w(\mathbf{n}, \nabla\mathbf{n})dx,$$

with Ω a give domain in $\mathbf{R}^3$, $\mathbf{n}$ a vector field, constrained to be of unit magnitude,

$$\mathbf{n}\cdot\mathbf{n} = 1,$$

with Lagrangians of the form

$$2w(\mathbf{n}, \nabla\mathbf{n}) = K_1(\nabla\mathbf{n})^2 + K_2(\mathbf{n}\cdot \text{ curl } \mathbf{n})^2 \\ + K_3 \mid \mathbf{n}\times \text{ curl } \mathbf{n} \mid^2 + \alpha[\text{tr}\nabla\mathbf{n}^2 - (\nabla\mathbf{n})^2].$$

Here, the K's and α are constants, restricted so that w is non-negative. The special case

$$K_1 = K_2 = K_3 = \alpha = K \Rightarrow 2w = K \mid \nabla\mathbf{n} \mid^2$$

also arises in the theory of harmonic mappings. This has produced some fruitful exchanges of ideas, although various serious difficulties have been encountered, in trying to adapt analyses to more general forms of w. Boundary conditions of physical interest do include those of the familiar Dirichlet type. Then, minimizers must exhibit singularities at points in Ω if the boundary values are smooth, but not of topological degree zero, and may do so if this degree is zero. Discussion of this topic will follow rather closely the survey by Cohen, et al., in reference [**1**].

Other mathematical topics to be discussed, and more, will be covered in summary articles by Kinderlehrer [**2**] and Lin [**3**]. For one thing, other boundary conditions of physical interest leave **n** free to take on any values consistent with the condition that it makes a fixed angle with the normal to $\partial\Omega$. This tends to produce singularities at points on $\partial\Omega$, possibly also in Ω; existence and regularity theory have been extended to include such problems. Also, theory sketched above is not very satisfactory for dealing with some observed line singularities or defects. The lecture will touch upon ideas used to modify theory to correct this, and some results which have been obtained, using it.

References

1. "Theory and Applications of Liquid Crystals," (ed. J. L. Ericksen & D. Kinderlehrer), Institute for Mathematics and Its Applications Volumes, Vol. 5, Springer-Verlag, 1987.
2. D. Kinderlehrer, *Recent developments in liquid crystal theory*, to appear in Proc. Colloque Lions, Paris (1988), to be published by North-Holland.
3. F.-H. Lin, *Nonlinear theory of defects in nematic liquid crystals—phase transition and flow phenomena*, to appear in Comm. Pure Appl. Math.

Department of Aerospace Engineering and Mechanics and School of Mathematics, University of Minnesota, Minneapolis, MN 55455

The Etruscan Venus

GEORGE K. FRANCIS

Introduction.

State of the art computers, such as the Silicon Graphics IRIS, have something to offer journeyman topologists that can otherwise be experienced only in the imagination. It is the wonder of animating, in real time, complicated deformations of topological surfaces, interactively! This paper reports a project at the National Center for Supercomputing Applications (NCSA) to visualize a regular homotopy of a Klein bottle immersed in 4-space. The shadow (projection) of this phenomenon in 3-space is a mapping homotopy between stable images of closed, one-sided (non-orientable) 2-manifolds called *ovalesques.* Such surfaces are generated by the prescribed motion of an oval (e.g. an ellipse) through space. Thus, the notion of an ovalesque is a projective generalization of a ruled surface. Recall that ruled surfaces are generated by straight lines.

The initial results were announced in [**5**]. A popular account of the project by Ivars Peterson [**11**] will also be included in his forthcoming book. Computer generated pictures can be very beautiful. The aesthetics of the project are elaborated by Donna Cox [**4**]. They were also discussed in the author's contribution to the splendidly illustrated Encicolopedia Italiana volume produced by Michele Emmer [**8**]. Technical aspects of the custom rendering software by Ray Idaszak [**9**] were incorporated in an exhibit at the Chicago Museum of Science and Industry [**3**], and in Ellen Sandor's phscolograms [**12**]. A 3.5 minute narrated videotape is part of Maxine Brown's 1889 SIGGRAPH tape [**6**]. A general introduction to the descriptive topology which provides the mathematical basis of the project may be found in Chapter 5 and the Appendix of *A Topological Picturebook* [**7**].

The Romboy Homotopy Project.

The *Venus* is a topological Kleinbottle, mapped into 3-space under a stable (generic) projection from an embedding in 4-space. As such, she joins similar and more familiar surfaces: Steiner's *crosscap* and *Roman* surfaces, and the surface of Werner Boy. The Roman surface is the 3-dimensional shadow of a projective plane embedded in 4-space, called the *Veronese* surface. Topologically speaking, a *projective plane* is a Moebius band glued edge-to-edge to a disc. A Kleinbottle consists of two Moebius bands glued edge-to-edge. The Venus and the Roman surface have, in addition to double lines, also pinchpoint singularities. The reader is directed to a thorough treatment of these matters in the *Topological Picturebook* [7], especially Chapter 5. Boy's surface has no such singularities, it is said to be immersed in 3-space. The ***Romboy Homotopy*** is a deformation of the Roman surface into Boy's surface. This deformation was parametrized and studied extensively by François Apéry [1] in his dissertation under the great topological visualizer, Bernard Morin of Strasbourg.

We report here on a project at the National Center for Supercomputing Application (NCSA), under the direction of Donna Cox [4], the collaboration of the author, and programmed by Ray Idaszak [9] , to visualize the Romboy homotopy. In the process we developed a simplification of the Roman surface to the Venus and a generalization of the homotopy which desingularizes the Venus. The homotopy produces a Kleinbottle immersed in 3-space named *Ida*. Its namesake, Ray Idaszak, was the first to see this new surface on the computer screen. Color Plate 19 shows a Venus/Ida pair just before and just after the pinchpoint cancellation stage. Color Plate 20 shows the Venus at the initial setting of all parameters and extended in 4-space. The fourth coordinate is painted on the surface using cyclic color tables at two different frequencies. Color Plate 21 is a stereogram of the familiar 4-dimensional surface, Steiner's crosscap. Color Plate 22 is a shaded stereogram that shows how the Venus is the connected sum of two Roman surfaces. The use of stereopairs and surface color to assist visualizing 3- and 4-dimensional surfaces is discussed in the second section.

For his purposes, Apéry chose a special parametrization of the Roman surface as a bouquet of ellipses, all of which are tangent to the horizontal plane at the origin. Such a surface is an *ovalesque* since it is swept out by a moving oval, see Figure 1. Geometrically, we have a map of a torus into space. During the homotopy each oval is stretched within its own plane. The ellipses of Apéry's ovalesques are doubly covered by their meridinal parameter. There is a deformation of such a circle, through the *limaçons* of Pascal, to a larger circle. Uncurling the the ovals of the Roman surface through the limaçons produces a more primitive, "pre-Roman" ovalesque

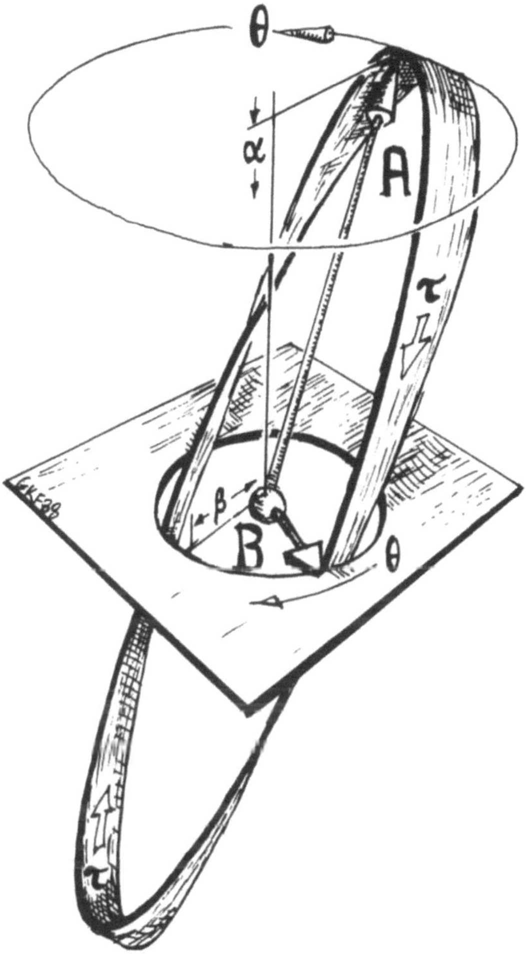

Figure 1: Ovalesque

which, rendered and viewed just right, is aptly called the *Etruscan Venus*.

At heart of every computer program for drawing an ovalesque is a very simple algorithm, nicknamed ETR, which generates the desired number of vertices and surface normals (when needed) so efficiently that we have found it unnecessary to use the customary display list definition for geometric objects. Indeed, the preliminary work for this project was performed on an Apple IIGS using Robert Illyes' FORTH compiler **[10]**. The efficiency of the ETR algorithm permitted Illyes to implement it on an experimental graphics computer using a Novix 4000 chip. This 16 bit 4 mHz RISC chip, whose machine language is FORTH, can execute in excess of one instruction per cycle, or roughly 4 high level MIPS. The chip generates a video signal directly, without an interface chip. At 200 lines per frame there is enough time for the chip to calculate the position, select a Lambert shade, and Z-buffer each point of the stereogram, Color Plate 22, in ca. 45 seconds. That is about the same time it takes to render the surface on an older IRIS workstation, albeit in fine color. The depth-cued wire-frame runs at two frames per second. Thus an almost acceptable animation can be achieved by running the chip at a much higher clock rate.

To explain the ETR algorithm, let me assume that you have some way of rendering a sphere or torus on your favorite graphics computer by writing this set of three parametric equations into your software, either as data or code.

$$\begin{aligned} &\text{For } 0 \le \beta \le \rho, \ 0 \le \alpha, \\ &\theta \in [0, 2\pi] \ \text{ and } \tau \in [0, 2\pi] \\ &x = \ \ (\rho + \beta \sin \tau) \cos \theta \\ &y = -(\rho + \beta \sin \tau) \sin \theta \\ &z = \ \ \alpha \cos \tau \end{aligned}$$

This *toroid* is swept out in a clockwise direction by an elliptical oval of height 2α and width 2β. If $\alpha = \beta$ and ρ shrinks to zero, a sphere of radius β is swept out by a great circle. Both of these surfaces, torus and sphere, are ovalesques. Let us rewrite the parametrization of an *ovoid* ($\rho = 0$) in two ways

$$\begin{bmatrix} 0 \\ 0 \\ \alpha \end{bmatrix} C_\tau + S_\tau \begin{bmatrix} \beta C_\theta \\ -\beta S_\theta \\ 0 \end{bmatrix} = [\, A_\theta \quad B_\theta \,] \begin{bmatrix} C_\tau \\ S_\tau \end{bmatrix}.$$

The first expression describes a vector pair (from the origin) which spans an ellipse as τ moves through its range, see Figure 1. The second expression presents the two conjugate axes of the ellipse as columns of a linear mapping of a circle into space. Suppose we tilt the A-axis and turn it counterclockwise at twice the speed of the equatorial B-axis.

$$A_\theta = \begin{bmatrix} C_{2\theta} \\ S_{2\theta} \\ \alpha \end{bmatrix} \qquad B_\theta = \beta \begin{bmatrix} C_\theta \\ -S_\theta \\ 0 \end{bmatrix}.$$

For $\alpha = \sqrt{2}$ and $\beta = 1$, approximately, this parametrizes the Etruscan Venus. The surface is a singular Kleinbottle, topologically equivalent to the connected sum of two copies of the Roman surface. Indeed, suppose we let the ovals curl up by multiplying the τ-circle by $\cos\tau$ thus

$$\begin{bmatrix} C_\tau \\ S_\tau \end{bmatrix} C_\tau = \begin{bmatrix} C_\tau^2 \\ C_\tau S_\tau \end{bmatrix} = \begin{bmatrix} \frac{1}{2} \\ 0 \end{bmatrix} + \frac{1}{2}\begin{bmatrix} C_{2\tau} \\ S_{2\tau} \end{bmatrix}.$$

Then for $\tau \in [0, \pi]$ and $\theta \in [0, 2\pi]$ we have Apéry's parametrization of the Roman Surface [1] . In the same work he computes a scalar divisor whose magnitude varies with both θ and τ.

$$1 - \frac{bS_{3\theta}S_{2\tau}}{\sqrt{2}}$$

It has the effect of magnifying and tilting the generating ellipse of the Roman surface in the plane determined by the conjugate axes A and B. With increasing b the singularities of the Roman surface cancel pairwise, at $b = 1/\sqrt{3}$. The result is Boy's surface. The deformation is the ***Romboy Homotopy*** and may be parametrized thus

For $\theta \in [0, \pi]$ and $\tau \in [0, 2\pi]$

$$\begin{bmatrix} x \\ y \\ z \end{bmatrix} = \frac{(1-\ell) + \ell C_\tau}{1 - bS_{3\theta}S_{2\tau}/\sqrt{2}} \begin{bmatrix} C_{2\theta} & \beta C_\theta \\ S_{2\theta} & -\beta S_\theta \\ \alpha & 0 \end{bmatrix} \begin{bmatrix} C_\tau \\ S_\tau \end{bmatrix}.$$

Note the scalar factor, whose denominator is the Romboy homotopy and whose numerator is the limaçonic homotopy. The latter has polar coordinates

$$r = (1-\ell) + \ell\cos\tau.$$

During $0 < \ell < 1$ the surface is not, strictly speaking, an ovalesque, unless we wish to think of the limaçons as honorary ovals. Apéry's homotopy applied to the Venus ($\ell = 0, b = 0$) also cancels the pinchpoints (12 in number) to produce the immersion Ida. The monochrome pair on Color Plate 19 shows Venus and Ida just before and after the cancellation. Ida is, topologically, the connected sum of two copies of Boy's surface.

Painting the 4th Dimension.

How shall we train our vision to integrate systematically encoded information about higher dimensional objects and events in a way which is analogous to the way we routinely recognize 3D information in a flat picture? That is a central problem in scientific visualization. The task of descriptive topology and computer geometry is to solve it. The historical precedent is, of course, the invention and codification of linear and aerial perspective half a millenium ago. Mathematics joined with Art in the Rennaissance to solve this picturing problem. For a century and a half the geometry of Gauss and Riemann has challenged mathematics teachers, model builders, and progréssive artists to devise concrete realizations of what is, in the abstract, a very small induction step from 2 to 3 to 4 to ... . The computer based geometry films of Banchoff and Strauss **[2]** are the pioneering works in this program. Simple surfaces in 4-space rotate rigidly while being projected into 3-space. There we see their shadows as wire-frame surfaces which undergo initially strange but eventually recognizable and ultimately familiar deformations. The models turn slowly in 3-space to help the 3-D visualization from the picture screen.

Another method is to simply slice the 4-D object into a succession of 3-D sections transverse to a fixed direction. The temporal succession of these slices yields an animated representation of the object, in which time is the fourth dimension. This method precludes using time again to depict a 4-dimensional homotopy.

In our project we decided to explore using color to encode additional dimensions. For this purpose one needs an accurate way of handling computer color. The color mixer I am about to describe is small enough to fit right into the ETR program, as a subroutine. It is also versatile enough to approximate most of the palettes and coloring schemes we experimented with, but which were produced by various other methods.

To each point on the surface there is an additional scalar value t (for "temperature", not for "time") which we plan to interpret as a color. For example, we may write the crosscap as an ovalesque, see Figure 2, by projecting this embedding of the projective plane in 4-space to the first 3 coordinates of

$$\begin{bmatrix} x \\ y \\ z \\ t \end{bmatrix} = \begin{bmatrix} C_{2\theta} \\ S_{2\theta} \\ 0 \\ 0 \end{bmatrix} C_\theta^2 + S_\tau C_\tau \begin{bmatrix} 0 \\ 0 \\ C_\theta \\ S_\theta \end{bmatrix}.$$

For easier visualization of the crosscap, Color Plate 21, map the t-coordinate into a five color scale:

(−0.5) blue (−0.3) cyan (−0.1) green (+0.1) yellow (+0.3) red (+0.5).

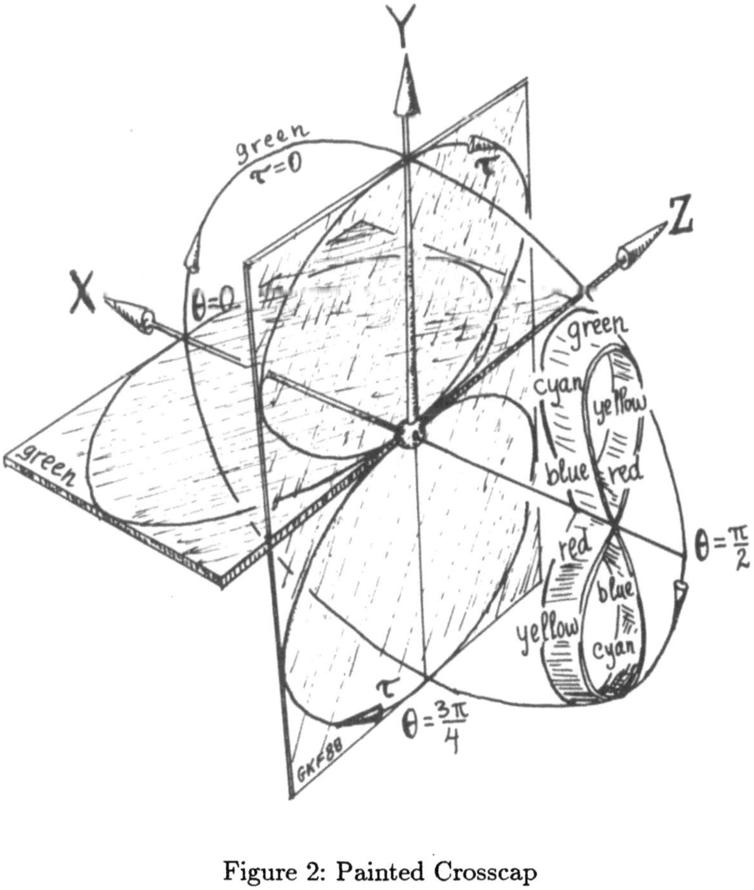

Figure 2: Painted Crosscap

Note that along the double line the two sheets never cross on the same color. This verifies that the above parametrizes an embedding of the cross-cap in 4-space. In general, t is mapped into a linear (or circular) palette given by three intensities of red, green and blue, $(r(t), g(t), b(t))$, call it the *paint* on that point. It is necessary to distinguish the *paint* from the actual *color* sent to the corresponding pixel because the latter also includes information about the illumination. Idaszak showed that the following illumination and shading model was perfectly adequate for our purposes.

Let λ denote the absolute value of the Lambert cosine. In Lambert's cosine law, cf Chapter 3 of [7], only the angle between the light source and the surface normal is significant. Since our surfaces may be non-orientable, it is appropriate to use a projective (non-oriented) normal. Consequently, portions of the surface facing away from the light source are too brightly illuminated. But this is a small price to pay. There are, in fact, several bonuses. For one, you should place the light source behind you and not too far to one side. That way most inappropriately illuminated spots on the surface are invisible. Moreover, it is a regular painterly device to highlight a narrow portion between the apparent contour and the shadow boundary of a curved surface (see Dali's paintings). Our "projective" illumination artifact helps a little in this direction, see Color Plates 19,22. On more extensive portions of the surface in the shadow, it provides backlighting (albeit too much) for no extra cost. *Backlighting* is a good cure for the "dimpling" rendering error which occurs when the shaded portion of an (oriented) surface is given a flat ambient value of its color. This optical illusion, related to the Mach band phenomenon, tends to alter the apparent curvature of the surface, introducing negative curvature where there is none.

The precise composition of the palette is of utmost importance, as Cox has shown in her many scientific visualization projects at the NCSA. Let $p(t)$ denote one of the paint functions, $r(t), g(t)$ or $b(t)$. Each of these shall be a truncated sinusoidal wave

$$p(t) = \operatorname{trunc}_0^1 [\, \beta + \alpha \cos(\lambda t - \phi)\pi \,] .$$

Thus each primary is given a baseline β, amplitude α, frequency λ, and phaseshift ϕ. See Figure 3. For simplicity we took the ambient and the specular illumination from a gray scale. For this, two further parameters suffice, μ and σ. μ is for the ambient value and σ insures that as we

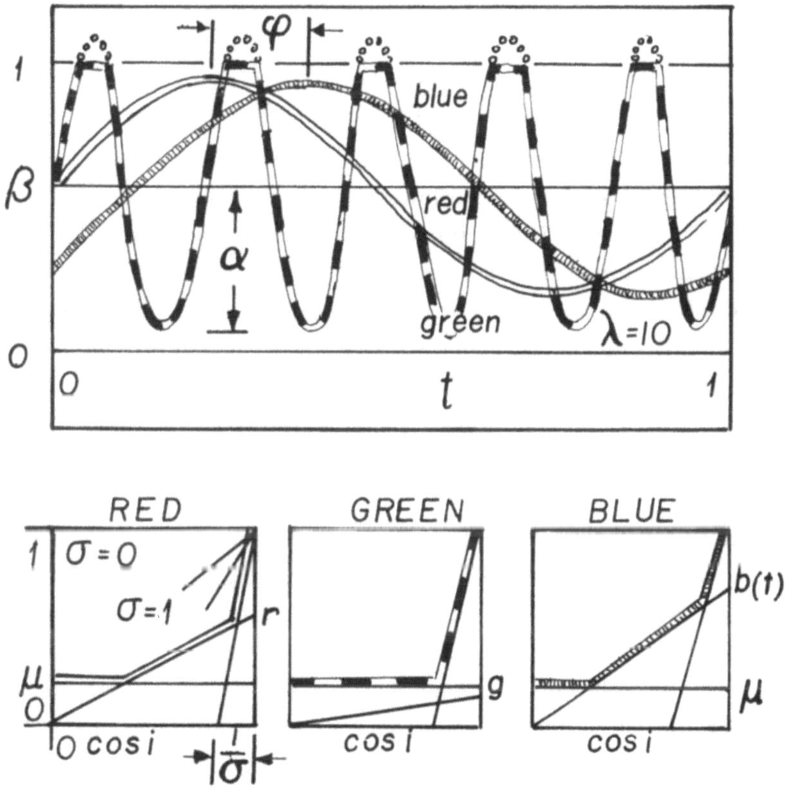

Figure 3: Cosine Colormixer

approach the ***brilliant point*** (=maximal Lambert cosine) the color rises steeply to white. Nothing is gained by using floating point exponentiation. Use the linear function

$$s(t) = (x - \frac{1}{\sigma})(\frac{\sigma}{\sigma - 1}).$$

in assigning the final color for a pixel thus

$$\begin{bmatrix} R(t) \\ G(t) \\ B(t) \end{bmatrix} = \begin{bmatrix} \max\{\mu, r(t), s(t)\} \\ \max\{\mu, g(t), s(t)\} \\ \max\{\mu, b(t), s(t)\} \end{bmatrix}.$$

To check how the Venus could be the shadow from hyperspace, consider an embedding whose fourth coordinate is just $w = \sin\theta \sin\tau$. For the left image in Color Plate 20 this function is mapped once around a circular palette with magenta the "wrap-around" indicator. Note that along the visible double lines the colors separate convincingly except, perhaps, near the upper left pinchpoint. Quadrupling the frequency of the color map for the right image resolves this uncertainty. On the other hand, newly suspicious matches along the green paint on the right image do not signify self-intersections. That may be checked on the "slower" color map on at the left.

A particularly interesting experiment is to use a very high frequency color table. Past the Nyquist limit, where sampling frequency drops below half the signal frequency, *aliasing* occurs. The effect on the surface paint is that spurious extrema and saddles are marked on the surface. It is difficult to distinguish the real from the apparent critical points in a still picture. In the animation **[6]**, where the surface undergoes a slow 4-dimensional deformation, the apparent critical points sparkle in a way that easily distinguishes them from the real extremata. In situations where it is impractical to scale a function defined on a surface, the use of high high frequency color maps is a way of estimating its Morse-theoretic properties visually. This is the subject of a future study.

REFERENCES

1. François Apéry, "Models of the Real Projective Plane," F. Vieweg Sohn, Braunschweig/Wiesbaden, 1987.
2. Thomas Banchoff and Charles Strauss, "Complex Function Graphs, Dupin Cyclides, Gauss Map, and Veronese Surface Computer Geometry Films," Brown University, Providence, 1978.
3. Maxine Brown, *The interactive image*, IEEE Comp. Graphics Applic., January (1988).
4. Donna J. Cox, *Using the supercomputer to visualize higher dimensions*, Leonardo **21** (1988).
5. George K. Francis, Donna J. Cox, and Ray L. Idaszak, *Romboy homotopy and Etruscan Venus: an experiment in computer topology*, Poster Session, SIAM Conference on Applied Geometry Albany, NY, July, 1987.
6. George K. Francis, Donna J. Cox, and Ray L. Idaszak, *The Etruscan Venus* (3.5 min narrated videotape in SIGGRAPH Video Review No. 49, T. DeFanti and M. Brown, editors), Assoc. Comp. Machinery **6** (1989).
7. George K. Francis, "A Topological Picturebook," Springer-Verlag, New York, 1987.
8. George K. Francis, *Topological art*, in "L'occhio di Horus," M. Emmer, ed., Enciclopedia Italiana, 1989.
9. Ray L. Idaszak, *A method for shading 3D single-sided surfaces*, IRIS Universe, Winter/Spring (1988), 9–11.
10. Robert F. Illyes, "ISYS Forth," Illyes Systems, Champaign, Illinois, 1988, 1984.
11. Ivars Peterson, *Twists of space*, Science News **32, no. 17** (1987), 264–266
12. D. J. Sandin, E. Sandor, W.T. Cunnally, M. Resch, T.A. DeFanti, and M.D. Brown, *Computer-generated barrier-strip autostereography*, Proc. Symp. Electronic Imaging, SPIE, January (1989).
13. Michael Ségard, *Art from the computer: an Illinois survey*, New Art Examiner, November (1988).

Department of Mathematics, University of Illinois, 1409 West Green Street, Urbana, IL 61801

Harmonic Mappings, Symmetry, and Computer Graphics

ROBERT GULLIVER

Abstract. A method for visually representing mappings from three dimensions into two is discussed and motivated by recent proofs that $x/|x|$ has minimum energy. Harmonicity is characterized in terms of the geometry of the representation. The angle problem for harmonic mappings is introduced.

Given two Riemannian manifolds M and N, of dimensions m and n respectively, we may define Dirichlet's integral, or the *energy*, of a mapping $u : M \to N$ as

$$E_2(u) = \int_M |\nabla u|^2 \, \mathrm{dVol}_M. \tag{1}$$

We assume N is embedded isometrically into Euclidean space $\mathbf{R}^d$ of some higher dimension d. A *harmonic mapping* $u : M \to N$ is one which is stationary for E_2 among mappings into N. The Euler–Lagrange equations are the elliptic system

$$\Delta_M u = \mathrm{tr}_M \, B\left(\frac{\partial u}{\partial x}, \frac{\partial u}{\partial x}\right), \tag{2}$$

where B is the vector-valued second fundamental form of N in $\mathbf{R}^d$, so that $B\left(\frac{\partial u}{\partial x_i}, \frac{\partial u}{\partial x_j}\right)$ is the component of $\frac{\partial^2 u}{\partial x_i \, \partial x_j}$ normal to N, and $\Delta_M = \mathrm{div}_M \, \mathrm{grad}_M$ is the Laplace–Beltrami operator of M.

If the exponent 2 in (1) is replaced by $1 \le p < \infty$, then one obtains the p-energy

$$E_p(u) = \int_M |\nabla u|^p \, \mathrm{dVol}_M.$$

Mappings which are stationary for E_p among all mappings: $M \to N$ satisfy the degenerate elliptic system

$$\mathrm{div}_M(|\nabla u|^{p-2} \, \mathrm{grad}_M \, u) = |\nabla u|^{p-2} \, \mathrm{tr}_M \, B\left(\frac{\partial u}{\partial x}, \frac{\partial u}{\partial x}\right). \tag{3}$$

Weak solutions of (3) are called *p-harmonic mappings.*

We are specifically interested in the case where M is the unit ball B^m in Euclidean m-space $\mathbf{R}^m$ and N is the unit n-sphere in $\mathbf{R}^{n+1}$. In this case, equation (2) may be written

$$(2') \qquad \sum_{i=1}^{m} \left[\frac{\partial^2 u}{\partial x_i^2} + \left| \frac{\partial u}{\partial x_i} \right|^2 u \right] = 0;$$

equation (3) is similarly simplified.

If M with its nonempty boundary is compact, then it is reasonable to choose a mapping $\varphi : \partial M \to N$ and to pose the boundary-value problem for equation (2) or (3) with Dirichlet boundary conditions:

$$(4) \qquad u = \varphi \quad \text{on} \quad \partial M.$$

Smooth, energy-minimizing solutions to the problem (2), (4) were found in 1964 by Eells and Sampson [**ES**] under the assumption that N has nonnegative sectional curvatures; and more generally in 1977 by Hildebrandt–Kaul–Widman [**HKW**] with the assumption that $\varphi(M)$ lies in a simple geodesic ball of radius $R < \pi/2b$ in N, where the sectional curvatures of N are bounded above by b^2. A new departure was the 1981 paper of Schoen and Uhlenbeck [**SU1**]. They allowed admissible mappings which were not continuous, and proved the existence of energy-minimizing maps without curvature conditions on N; under hypotheses weaker than those of [**HKW**], their solutions are smooth, but in general may have discontinuities on a set of codimension 3 (explicit examples below). An analogous theory for E_p with any $1 < p < \infty$ was carried out by Hardt and Lin [**HL**].

The theory of harmonic mappings holds strong analogies to the study of *minimal submanifolds.* In fact, the embedding of a minimal submanifold $M \subset N$ is harmonic in the induced metric; the analogies carry over to a much deeper level. The investigation of properties of two-dimensional minimal surfaces in a three-dimensional manifold has benefitted greatly from the intuition which is generated by visible models of minimal surfaces. Such models have been generated physically and examined for over a century: soap films, the interface of immiscible fluids at equal pressure, and even carved models. Recently, it has become possible to compute and display numerical approximations of minimal surfaces, as in the work of D. Hoffman and his collaborators. An analogous visualization of harmonic mappings is possible for the interesting case of mappings from 3-space into a 2-dimensional manifold, which we shall discuss in detail below.

The simplicity and naturality of the energy functional E_2 make it interesting from an abstract point of view, and also are responsible for its appearance in important applications. In *elasticity theory*, $E_2(u)$ may be seen as the energy of an isotropic and perfectly elastic material whose unstretched reference configuration is an infinitesimal homothety of M, after the material is deformed according to the mapping $u : M \to N$. A nematic *liquid crystal* with constant order parameter and with equal Frank constants $\kappa_1 = \kappa_2 = \kappa_3 > 0$ may be thought of as a harmonic mapping $u : \Omega \to S^2$ or $\mathbf{R}P^2$, with associated energy $E_2(u)$ ([**E**], [**HKL**]).

It is of interest to find explicit solutions of *smallest energy* for the boundary-value problem (2), (4) or (3), (4). A remarkably simple example of a harmonic mapping is the radial projection of the Euclidean unit ball B^m onto its boundary sphere S^{m-1}, namely $u_0(x) = x/|x|$, which has an isolated point of discontinuity, and has finite energy E_2 if $m \geq 3$. More generally, we may define a map $u_0 : B^m \to S^n$ for any $n \leq m-1$, by writing $x = (y, z)$ for $y \in \mathbf{R}^{n+1}$ and $z \in \mathbf{R}^{m-n-1}$ and defining $u_0(y, z) = y/|y|$: then u_0 has discontinuity along a subspace of codimension $n+1$, and has finite p-energy if $p < n+1$. One's first response is that something must be wrong when discontinuous solutions appear. In the context of elasticity theory, this model might be considered an oversimplification, since failure has occurred without leaving the elastic regime. In the context of liquid crystals, however, singularities are actually observed in the laboratory and must appear in some form with any usable model. Beginning with work of Jäger and Kaul in 1983 [**JK**] and continuing with results of Brézis–Coron–Lieb [**BCL**], of Lin [**L**] and our work with Coron [**CG**], it has been shown that the mapping u_0 is not only p-harmonic but even *minimizes* the p-energy:

THEOREM 1. *The mapping $u_0(y, z) = y/|y|$ from B^m to S^n has minimum p-energy for its boundary values for any integers m, n and p satisfying $1 \leq p \leq n \leq m-1$.*

We shall sketch the proof of Theorem 1, as given in [**CG**], for the special case $p = 2$. First suppose that the target dimension $n = 2$ as well. Then for any Lipschitz-continuous mapping $u : B^m \to S^2$, the *coarea formula* of Federer states that

$$\int_{B^m} J(u)dx = \int_{S^2} \mathcal{H}^{m-2}(u^{-1}(y))\ \mathrm{dVol}_{S^2}(y). \tag{5}$$

Here $J(u) = \sqrt{\det(\nabla u(\nabla u)^T)}$, and $\mathcal{H}^{m-2}$ denotes $(m-2)$-dimensional area, or Hausdorff measure. We need to apply the formula (5) to a dense subset of the space of mappings from B^m to S^2 with finite energy; although Lipschitz-continuous mappings are *not* dense for $m \geq 3$, there is a dense class $\mathcal{R}$ consisting of mappings which are locally Lipschitz continuous except on a well-behaved singular set of codimension 3, with controlled behavior near the singular set. For $u \in \mathcal{R}$, $u^{-1}(y)$ may have boundary in the interior B^m along the singular set of u, which would spoil the argument below; however, this is locally the same as the boundary of $u^{-1}(-y)$. That is, if we treat the points y and $-y$ in S^2 as having opposite orientation, then $u^{-1}(\{y, -y\})$ has its boundary only at ∂B^m. The advantage of this *fiber-balancing* technique will appear in inequality (6) below.

Now assume that $u \in \mathcal{R}$ has the same boundary values as $u_0 : B^m \to S^2$. Then $u^{-1}(\{y, -y\})$ has the same boundary as does $u_0^{-1}(\{y, -y\})$. Note that $u_0^{-1}(\{y, -y\})$ is the intersection with B^m of a codimension-2 subspace of $\mathbf{R}^m$. Therefore

$$(6) \qquad \mathcal{H}^{m-2}(u_0^{-1}\{y, -y\}) \leq \mathcal{H}^{m-2}(u^{-1}\{y - y\}) :$$

any flat disk has minimum area for its boundary. On the other hand, for any 2×2 matrix A, $2 \det A \leq |A|^2$, with equality only if A is an orientation-preserving similarity. It follows that $2J(u) \leq |\nabla u|^2$, with equality holding for $u = u_0$. Finally, formula (5) for u and for u_0 yields

$$\begin{aligned} E_2(u_0) &= 2\int_{B^m} J(u_0)dx = 2\int_{S^2} \mathcal{H}^{m-2}(u_0^{-1}(y))\, \mathrm{dVol}_{S^2} = \\ &= \int_{S^2} \mathcal{H}^{m-2}(u_0^{-1}\{y, -y\})\, \mathrm{dVol}_{S^2} \leq \int_{S^2} \mathcal{H}^{m-2}(u^{-1}\{y, -y\})\, \mathrm{dVol}_{S^2} = \\ &= 2\int_{B^m} J(u)dx \leq E_2(u). \end{aligned}$$

This proves Theorem 1 for the case $n = p = 2$.

For the general target dimension n, with $2 \leq n \leq m-1$, and with $p = 2$, the proof of Theorem 1 reduces by means of an *averaging property* to the case $n = 2$ just proved. In fact, consider any mapping $u : B^m \to S^n$ of finite energy. Let $\pi : \mathbf{R}^{n+1} \to \mathbf{R}^3$ be the projection onto the first three coordinates. For each rotation $Q \in \mathcal{O}(n+1)$, let the mapping $v_Q : B^m \to S^2$ be defined by $v_Q(x) = \frac{\pi \circ Q \circ u(x)}{|\pi \circ Q \circ u(x)|}$. Then for some constant c_n independent of u and of m, $c_n\, E_2(u)$ is the average over $Q \in \mathcal{O}(n+1)$ of $E_2(v_Q)$. Observe that if $u = u_0$, then $v_Q = v_{0Q} : B^m \to S^2$ is of the form $v_{0Q}(y, z) = y/|y|$, where $y \in \mathbf{R}^3$ and $z \in \mathbf{R}^{m-1}$, after a rotation of $\mathbf{R}^m$. Now suppose that

$u : B^m \to S^n$ has the same boundry values as $u_0 : B^m \to S^n$. Then for each $Q \in \mathcal{O}(n+1)$, v_Q has the same boundary values as v_{0Q}, so that $E_2(v_Q) \geq E_2(v_{0Q})$ by the case $n = 2$ considered above. Averaging over $\mathcal{O}(n+1)$ now yields $E_2(u) \geq E_2(u_0)$; this proves Theorem 1 in case $p = 2$.

A mapping $u : M^m \to N^n$, with $u \leq m$, is called *horizontally conformal* if, at each point of M, the restriction of ∇u to the orthogonal complement of its kernel is a similarity transformation. This property is related to the $n \times n$ Jacobian in the general form of the coarea formula (5), in the sense that $|\nabla u|^n \geq n^{n/2} J(u)$, with equality holding only where u is horizontally conformal. Theorem 1 is actually a special case of a more general theorem:

THEOREM 2 ([**CG**]). *Consider $u_0 : B^m \to S^n$ of finite p-energy, such that (1) u_0 is locally Lipschitz-continuous except on a piecewise smooth set of codimension $p+1$; (2) for almost all subspaces $Z \subset \mathbf{R}^{n+1}$ of codimension p, $u_0^{-1}(Z \cap S^n)$ minimizes $\mathcal{H}^{m-p}$ among oriented varieties with the same boundary; and (3) u_0 is horizontally conformal. Then u_0 minimizes p-energy for its boundary values.*

For the case $p = 2$, the proof of Theorem 2 is parallel to that of Theorem 1. When $p \neq 2$, the averaging property for E_p is no longer true, and must be replaced by an averaging property for the squares of the $p \times p$ Jacobians.

Three interesting examples of harmonic maps from B^{2n} to S^n are given by the homogeneous extensions of the *Hopf mappings*

$$u_0(z, w) = \frac{1}{|z|^2 + |w|^2}(|z|^2 - |w|^2, 2z \cdot \bar{w}) \in \mathbf{R} \times \mathbf{R}^n,$$

where $z, w \in \mathbf{R}^n$ and $z \cdot \bar{w}$ is the product of complex numbers ($n = 2$) quaternions ($n = 4$) or Cayley numbers ($n = 8$). These mappings satisfy the hypotheses of Theorem 2 with exponent $p = 2$ or $p = n$. Hypothesis (2) is satisfied for $p = n$, since $u_0^{-1}(\{s, -s\})$ is the union of two totally orthogonal n-planes in $\mathbf{R}^{2n}$. For $p = 2$, it may be shown that for any totally geodesic $S^{n-2} \subset S^n$, $u_0^{-1}(S^{n-2})$ is a complex quadric in $\mathbf{R}^{2n}$, with respect to an orthogonal complex structure which depends on the choice of S^{n-2}. This implies that $u_0^{-1}(S^{n-2})$ minimizes $\mathcal{H}^{2n-2}$ among oriented varieties with the same boundary.

A proof of Theorem 1 has been given by Avellaneda and Lin which is substantially simpler than the proof sketched above [**AL**]. In particular, they do not need to approximate mappings of finite energy by mappings with controlled singularities. For the case $u : B^m \to S^{m-1}$, their method utilizes the null-Lagrangian $I_p(\nabla u)$, the average of all $p \times p$ principal minors of the $m \times m$ matrix ∇u, which equals the average of products of p distinct eigenvalues of ∇u. This dependence on eigenvalues, rather than singular values, appears to be the only weakness of their method: Theorem 2 (when $n \geq 3$), and in particular the minimizing property of the Cayley and quaternionic Hopf maps, do not appear to follow from their proof. However, it seems likely that their method may be extended significantly by considering a null-Lagrangian constructed in terms of a specific harmonic mapping u_0.

Harmonic mappings $u : B^m \to S^{m-1}$ may be interpreted in another way, as harmonic *unit vector fields* on B^m. This is especially appropriate in the theory of liquid crystals. This point of view suggests that harmonic mappings $u : M^m \to N^n$ from one Riemannian manifold to another might usefully be replaced by the more general context of *harmonic sections* of a Riemannian fiber bundle.

On a more concrete level, Theorem 1 (with $p = 2$ and $n = m - 1$) may be interpreted as stating that the unit radial vector field on B^m has the smallest energy among unit vector fields which are normal at the boundary of B^m. A natural problem now arises, which we call the *angle problem.* Consider unit vector fields on B^m which form a prescribed (constant) angle α with the exterior normal vector at each boundary point of B^m; among such vector fields, find one of smallest energy, and describe its geometry. Thus the Dirichlet condition (4) is replaced by the partially free boundary condition

$$(4') \qquad u(x) \cdot x = \cos\alpha \quad \text{whenever} \quad |x| = 1.$$

We expect that an abstract existence theorem can be proved, along the lines of [**SU2**]; regularity at the boundary will necessarily be weaker than the conclusions of [**GJ**]. In fact, if $0 < \alpha < \pi$ and m is odd, then the solution *must* have boundary singularities, since its tangential component is a vector field on S^{m-1} and cannot be continuous. Moreover, unlike the $\alpha = 0$ case, there is no *topological* reason to require interior singularities.

For $m = 3$ and α close to zero, one may show that any minimizer for the angle problem has a singularity close to the origin, using the uniqueness of $u_0(x) = x/|x|$ ([**BCL**], Theorem 7.1). Several questions now arise. Must this interior singularity persist for all α? For $m = 3$, a boundary singularity has fractional degree $\pm\frac{1}{2}(1-\cos\alpha)$ plus an arbitrary integer. What values of the degree may occur at boundary singularities of a minimizer? How many boundary singularities occur? Where are the boundary singularities located if, as we conjecture, there are exactly two? Observe that for $\alpha = 0$, the unique minimizer $u_0(x) = x/|x|$ is $\mathcal{O}(3)$-equivariant. In contrast, no $\mathcal{O}(3)$-equivariant vector field satisfies the boundary condition for $0 < \alpha < \pi$; the highest symmetry possible would be $\mathcal{O}(2)$. One is led to ask, is there any energy-minimizing solution for the angle problem which is equivariant for rotations about a fixed axis?

In the theory of harmonic mappings, we find that the angle problem is a specific problem of compelling simplicity. Moreover, the angle condition (4′) is a good model for certain liquid-crystal configurations. Yet the available mathematical techniques shed little light on questions such as those posed above. There is a need for sound conjectures to take the place of such questions. Numerical analysis and computer graphics may be useful in stimulating the correct intuition about the angle problem. Finite-element methods have been used to study the Dirichlet problem for unit vector fields on the three-dimensional cube, with the output in particular plane sections represented by arrows of varying length [**CHKLL**]. In the theorems above, we have seen the importance of the length of fibers $u^{-1}(y)$ or $u^{-1}(\{y,-y\})$ and the deviation from horizontal conformality for minimizing properties of maps $u : B^3 \to S^2$. A useful *visual representation* of an energy-minimizing map $u : B^3 \to S^2$ should therefore be constructed by choosing an approximately uniformly distributed set of points $\{y_1, \cdots, y_n\}$ on S^2 and drawing the inverse image of the circles of a small radius ϵ about these n points. This inverse image would consist of n thin, smooth tubes of elliptical cross-section in the three-dimensional ball. All n tubes would converge to any interior singularity, becoming thinner and more circular in proportion to the distance from the singularity. A subset of the tubes would converge to each boundary singularity. Elsewhere, the tubes would balance three opposing forces, which may be described in terms of the geometry of the tube as follows. Let $\lambda_1 \geq \lambda_2 \geq 0$ be the singular values of ∇u, so that the elliptical cross-section of the tube has major axis $2\epsilon/\lambda_2$ and minor axis $2\epsilon/\lambda_1$. Let k_3 be the curvature vector of the central curve of the tube and let k_2 and k_1 be the curvature vectors of the orbits of the major and minor axes. Then in the direction of the minor axis, the total force acting on the tube is the orthogonal component of

$$(2'') \qquad \operatorname{grad}(\lambda_1^2 - \lambda_2^2) - 2(\lambda_1^2 - \lambda_2^2)k_2 - 2\lambda_1^2 k_3.$$

In the direction of the major axis, the subscripts 1 and 2 are exchanged.

The problem of effective visual representation of a mapping $u : B^3 \to S^2$ is thereby reduced to the problem of interpreting a two-dimensional image of a smooth surface (the tubes). The latter problem has been the subject of considerable work in recent years, and those methods which have proved successful could profitably be incorporated to enhance our understanding of harmonic mappings. It is to be hoped that well-conceived graphics, combined with fast numerical algorithms, will allow investigators to exploit the powerful parallel processor which is the human visual facility in developing their understanding of nonlinear mathematical models.

We would like to thank J.–M. Coron and S. Luckhaus for stimulating conversations, and the C.M.A. at Canberra for its hospitality.

REFERENCES

[AL] Avellaneda, M. and Lin, F.-H., *Fonctions quasi-affines et minimization de* $\int |\nabla u|^p$, C. R. Acad. Sci. Paris **306**-I (1988), 355–358.

[BCL] Brézis, H.; Coron, J.-M. and Lieb, E., *Harmonic maps with defects*, Commun. Math. Phys. **107** (1986), 649–705.

[CG] Coron, J.-M. and Gulliver, R., *Minimizing p-harmonic maps into spheres*, J. Reine Angew. Math. (to appear).

[CHKLL] Cohen, R.; Hardt, R.; Kinderlehrer, D.; Lin, S.-Y. and Luskin, M., *Minimum energy for liquid crystals: computational results*, IMA Preprint #250, Minneapolis (1986).

[E] Ericksen, J. L., *Equilibrium theory of liquid crystals*, Advances in Liquid Crystals, Vol. 2, Brown, G. H. (ed.), Academic Press, New York (1976), 233–299.

[ES] Eells, J. and Sampson, J. H., *Harmonic mappings of Riemannian manifolds*, Amer. J. Math. **86** (1964), 109–160.

[GJ] Gulliver, R. and Jost, J., *Harmonic maps which solve a free-boundary problem*, J. Reine Angew. Math. **381** (1987), 61–89.

[HKL] Hardt, R.; Kinderlehrer, D. and Lin, F.-H., *Existence and partial regularity of static liquid crystal configurations*, Commun. Math. Phys. **105** (1986), 547–570.

[HL] Hardt, R. and Lin, F.-H., *Mappings minimizing the* L^p *norm of the gradient*, Comm. Pure Appl. Math. **40** (1987), 555–588.

[HKW] Hildebrandt, S.; Kaul, H. and Widman, K.-O., *An existence theory for harmonic mappings of Riemannian manifolds*, Acta Math. **138** (1977), 1–16.

[JK] Jäger, W. and Kaul, H., *Rotationally symmetric harmonic maps from a ball into a sphere and the regularity problem for weak solutions of elliptic systems*, J. Reine Angew. Math. **343** (1983), 146–161.

[L] Lin, F.-H., *A remark on the map* $x/|x|$, C.R. Acad. Sci. Paris **305-I** (1987), 529–531.

[SU1] Schoen, R. and Uhlenbeck, K., *A regularity theory for harmonic maps*, J. Diff. Geom. **17** (1982), 307–335.

[SU2] ——————, *Boundary regularity and the Dirichlet problem for harmonic maps*, J. Diff. Geom. **18** (1983), 253–268.

Centre for Mathematical Analysis, Australian National University, Canberra ACT 2601, Australia and School of Mathematics, University of Minnesota, Minneapolis, MN 55455 U.S.A.

Video Based Scientific Visualization

W. E. JOHNSTON, D. W. ROBERTSON, D. E. HALL,
J. HUANG, F. RENEMA, M. RIBLE, AND J. A. SETHIAN[1]

Abstract. We describe techniques for producing video imagery for scientific visualization. These techniques involve variations of graphics algorithms, distributed computing, and a versatile, low cost, video movie making system. Although video can be used for single frame displays, its obvious advantage is for animated movies. Video movies are made by single frame animation from the output of modeling processes like time dependent, numerical simulations. Visualization algorithms are used to convert abstract data into a geometric form for graphical display. The system uses a distributed architecture and extensive data compression to permit the use of wide area, as well as local area networks connecting the systems generating the data, the graphics, and doing the video recording.

1. Introduction.

Visualization of abstract data is an important aspect of understanding that data. Time dependent data may often be displayed as animated sequences that we will refer to as "scientific movies". These movies provide a useful part of visualizing data such as flow fields and moving surfaces, and for complex 3D structures where animation by rotation and translation (instead of by time dependency) provides necessary visual clues. However several issues need to be addressed in order to make sophisticated visualization available to the scientific users that have the most need for this methodology. Beyond the questions of the appropriate visualization techniques for a particular type of data, and the graphics support needed to generate the image, are the key areas of designing the user interface and building software that will transparently make use of the necessary resources.

[1]E-mail addresses are: wejohnston@lbl.gov, or ...ucbvax!lbl-csam.arpa!johnston; dwrobertson@lbl.gov, hall@casm.lbl.gov, huang@csam.lbl.gov, max@csam.lbl.gov, sethian@csam.lbl.gov

The work presented in this paper is supported by the Director, Office of Energy Research, Scientific Computing Staff, Energy Sciences Advanced Computing Program, of the U.S. Department of Energy under contract DE-AC03-76SF00098. Any conclusions or opinions, or implied approval or disapproval of a company or product name are solely those of the authors and not necessarily those of The Regents of the University of California, the Lawrence Berkeley Laboratory, or the U.S. Department of Energy. Trademarks are acknowledged by TM.

In this paper we will discuss several aspects of a system that tries to address the issues mentioned above. A "typical" visualization algorithm is presented, together with its relationship to both the user interface above it, and the graphics system below it. We then briefly examine the issues in using a distributed system to access and manage the resources needed to make "video movies". Throughout the paper we will use the scenario of a modeling process running on a supercomputer, the visualization and resulting graphics being generated at a local workstation, and the video recording being done at another workstation. Although there is nothing in the system that dictates this distribution of the software modules, our implementation most frequently runs in this environment.

2. Video Based Visualization.

Video graphics are different from pen plotter, terminal and film graphics. The spatial resolution of the encoded video signal[2] used for video recording is low compared to the more traditional media, while the color resolution is relatively high [**1**] and [**2**]. These factors must be taken into account when designing a visualization to be displayed using video tape. One does not count on producing the fine line drawings that are displayed, for instance, on laser printers, but rather uses smoothly shaded areas that are best reproduced on video media. However, even with the relatively low spatial resolution of NTSC video, line and point drawings can still be useful if special care is taken in their design. Color Plate 25 shows sample frames from two movies made by using this system for visualizing supercomputer generated data, by A.F. Ghoniem and J.A. Sethian.

The system provides interfaces at several levels, including the compression and transport of a frame buffer, 2D and 3D graphics primitives, and visualization modules such as 3D surface tessellation and particle advection.

The design goal of the system is to provide scientific visualization using minimal implementations of graphics rendering[3] algorithms. The minimality issue is driven by the desire to avoid the complexity and expense of

[2] The basic principles for encoding the red, green and blue color intensities that come out of a typical hardware frame buffer into a single signal that can be used for tape recording and broadcast are similar for all three commonly used standards—NTSC, PAL and SECAM.

[3] The term "render" is used in computer graphics to describe the process by which graphics entities are converted to a raster image. This process is responsible for surface shading, lighting, etc.

graphics rendering algorithms needed to produce realistic images, versus the need for enough visual realism to resolve any viewing hypothesis ambiguities. A certain degree of "realism" is necessary, for instance, when presenting planar projections of 3D geometric shapes in order to ensure that the observer can accurately reconstruct the original 3D spatial relationships, and when rendering the interior of a hollow object in order to be able to recognize that we are looking at the concave inner surface of the object, rather than the convex outside surface. The use of simple "flat" shading, for instance, frequently produces an image in which the concave-convex distinction cannot be made. On the other hand, for the Kummer surface[4] in Color Plate 26, obtained by Elaine Yeh, W.E. Johnston, D.W. Robertson, and M. Rible, the function is evaluated on a $100 \times 100 \times 100$ grid, and the tessellated surface for one function value results in about 100,000 polygons. Flat shading is generally used with this many polygons, as more sophisticated shading is both expensive and not needed to improve the visualization. Our approach is to use the minimal shading and lighting algorithms necessary to accomplish an unambiguous representation. This reduces CPU and network demands, and keeps the software manageable. See [**3**] for more information.

3. A Distributed System for Video Based Visualization.

Our implementation of a video based visualization system is a distributed application with several modules. These software modules are designed to be distributed among heterogeneous systems, since the resources required by the various modules frequently do not reside on a single system. This is especially true for the video signal generation and recording.

[4] $f(x,y,z) = x^4 + y^4 + x^4 - (y^2z^2 + z^2x^2 + x^2y^2) - (x^2 + y^2 + z^2)$, see Fischer [**4**].

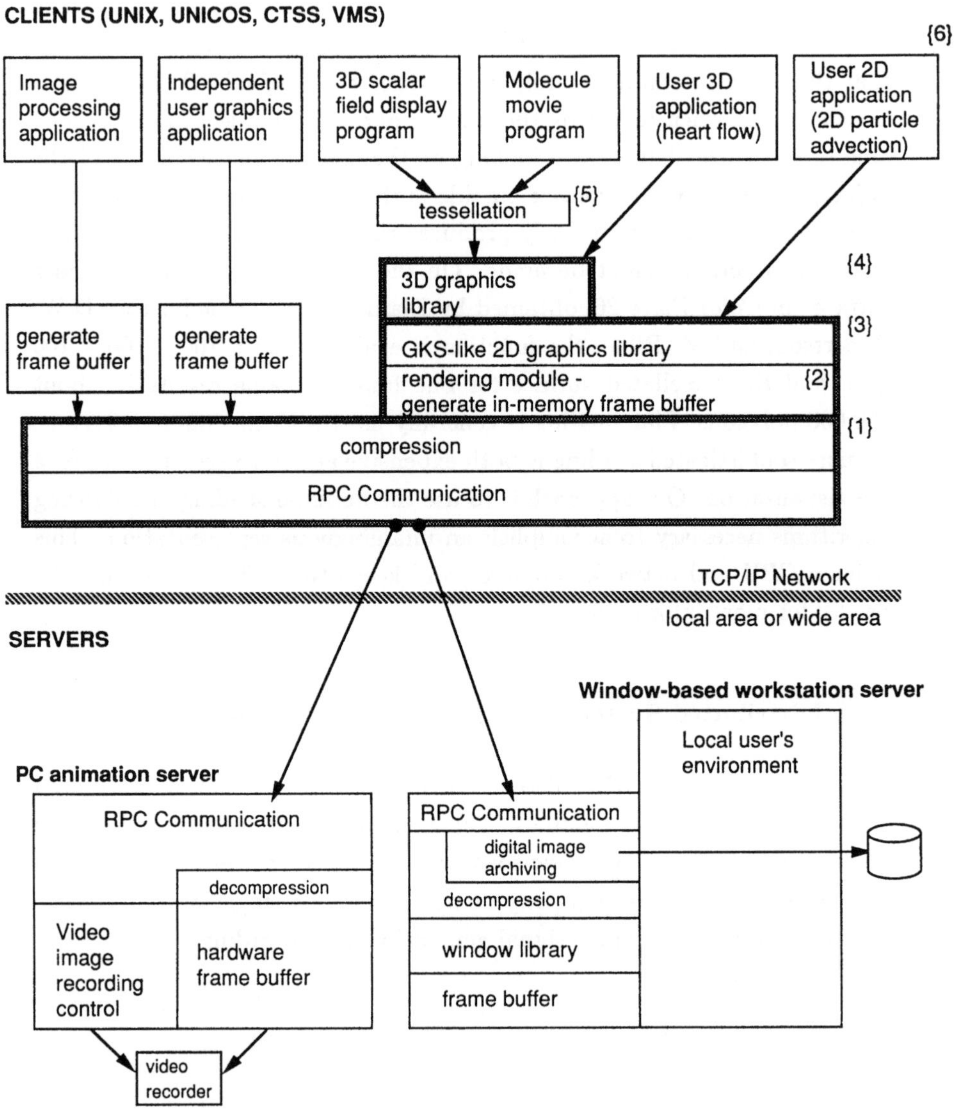

Figure 2 - Software architecture of Scry—a distributed imaging and visualization system

The modules that make up the system are illustrated in Figure 2. There are four main groups of modules:

(1) Generation of a numerical representation of some mathematical model (the application),
(2) conversion of the numerical representation to a graphics representation (the visualization),
(3) conversion of the graphics representation to a raster image representation (the rendering), and
(4) the display and recording of the image.

These steps are connected procedurally on a local system, or by an inter-process communication mechanism when the modules reside on different systems [**5**].

4. The Numerical Representation: Generating the Data.

The supercomputer application is a numerical implementation of a mathematical model. Typical applications include modeling of 2D and 3D flow fields, wave propagation, 3D surface topology studies, and various physics simulations (e.g. accelerator design studies) that produce 2D and 3D particle position fields.

5. The Visualization: Conversion to Graphics Primitives.

The models produce numerical data for which techniques must be developed in order to visualize the information content of that data. Sometimes these techniques are obvious, and sometimes it is difficult to design a pictorial representation that will adequately display the important aspects of the data. Visualization algorithms (layers {5} and {6} of Figure 2) convert a numeric representation of the model output into the graphics primitives (2 and 3D points, lines, polygons and text) that are used in a geometrical representation. The visualization process is not graphics *per se*, but algorithms that transform the numeric representation to a geometric form that can be displayed as an image. Typical examples of visualization algorithms are: the construction of the level curves through a 2D scalar field ($z = f(x, y)$) to produce "contour plots"; advection of particles through a flow field to obtain tracer particle positions that can be plotted (as in Color

Plate 25); and tessellation of the level surface of a 3D scalar field (Color Plate 26).

The graphics primitives produced by a visualization algorithm are sent to a rendering module for conversion to a 2D image.

5.1. 3D Surface Tessellation as a Visualization Technique.

As an example of a visualization technique, we consider how a 3D scalar field might be presented. One way to do this is to present the level surfaces of the scalar field. For existing rendering algorithms, one convenient representation of three-dimensional surfaces is by using covering polygons (patches). A cube can be represented by six squares, a tetrahedron by four triangles, and so on. For curved surfaces, it is necessary to use small planar polygons to approximate the curved surface. This does not necessarily produce a smooth surface: small spheres may look like faceted jewels rather than colored balls. There are, however, methods of shading (such as Gouraud shading) that blend the edges of polygons and leave the surface looking smooth again.

The computer representation of a 3D scalar field exists as values on a three-dimensional discrete grid. To tessellate a level surface one chooses a particular field value, and then looks for all places on the grid where neighboring grid point values straddle the chosen field value. This yields a representation of the intersection of the level surface with the grid. From here one approach to providing a tessellated surface is the "marching cubes" algorithm [**6**]. This algorithm works locally, seeking to fit a surface through each elemental cube (as defined by the the grid intersections) that intersects the surface. These elemental cubes have become known as "voxels" (volume pixels).

It can be shown that a maximum of four triangles are needed to represent the intersection of any planar surface with a cube. All possible intersections can be put into 256 classes, based on which corners of the cube are inside the surface and which are outside. Since being inside or outside is not significant, the number of different classes is reduced to 128, and these 128 are simply various rotations of 15 basic patterns. Thus, the "marching cubes" builds up a triangular tessellation of the entire surface by "marching" through the grid, one voxel at a time, and placing any triangles necessary to represent the surface's intersection with the grid in each voxel.

The result of this process is a collection of triangular polygons that represent the surface in question. While the triangles are of the same order of size as a linear dimension of a voxel, the placement of the triangle in the voxel is the result of interpolating between the grid points. The symmetries mentioned above ensure (for the most part) complete covering, and doing the same linear interpolation from voxel to voxel ensures that the triangles match at their edges. The triangles thus obtained are then given to a graphics rendering module for conversion to colored pixels that can be displayed on a 2D screen.

6. Graphics Rendering: Conversion to a Raster Image Representation.

For the purposes of this paper the output display is assumed to be a video image. These video images are generated from raster image data structures, which are scan line organized collections of pixels. Each pixel (elementary picture unit) is represented as a color value. The graphics rendering process converts the primitives generated by the visualization algorithm into a raster image. In our distributed movie system, graphics primitives are rendered by scan conversion into a software frame buffer (a data structure for storing raster images) located in main memory (level {3} in Figure 2). In the case of 3D primitives, hidden surface removal is performed by a z-buffer, also located in main memory. The software frame buffer is compressed and sent over the network to the video animation server. This occurs at {1} and {2} in Figure 2.

More information on rendering can be found in Rogers [7] and Foley and Van Dam [8]. It should be pointed out that not all graphics displays are the result of rendering traditional primitives. For example, "volume rendering" techniques produce images directly from 3D density (scalar field) information [9].

7. Numeric and Image Data Compression.

Once geometric primitives have been rendered into a raster image in the software frame buffer, that image needs to be displayed. Typically the rendering module runs as part of a "client" process on one system, and the display server is on another. When this is the case the raster image is compressed and sent over a network from the client to the server. Compression is especially important over wide area networks, where the observed bandwidth can be below 2K per second.

Compression can be divided into two types: entropy reduction, an irreversible compression since some (maybe insignificant) information is lost; and redundancy reduction, a reversible compression technique that tries to identify redundant information and encode it more efficiently. We use both types of compression, frequently together, in the movie system.

Synthetic raster images generated by computer graphics have an enormous amount of redundancy. There is typically high spatial coherence in the image, and high temporal (frame to frame) coherence. To compress these images, the system first applies entropy reduction techniques. Block truncation coding (BTC) [**10,11**] reduces the total number of colors needed to represent an image by reducing the colors in a block of pixels to two "best" representatives. A further compression is achieved by limiting the choice of possible representative colors to those of a short table whose index is then used to represent the colors in the blocks. Heckbert's median cut algorithm is used to populate the table with a set of colors that best reproduces the image [**12**]. One effect of this encoding is that areas of an image that appear different may result in the same block code, thus increasing the redundancy. The entropy reduced form of the image is then encoded with a redundancy reduction algorithm such as Lempel–Ziv coding [**13**].

Frame-to-frame differencing, which takes advantage of temporal coherency, can be applied before or instead of the Lempel–Ziv algorithm. In this case only the blocks of pixels that change are saved. Generally not much is gained by using both, since both take advantage of large groups of pixels being the same (which are less likely to change from one frame to the next). An exception is 2D images with a large amount of background, where the data structure describing which blocks of pixels change compresses very well using Lempel–Ziv. A disadvantage of frame-to-frame differencing is that it also requires that an additional frame buffer be allocated to hold the previous image.

Using these techniques in combination, the overall compression of synthetic raster images can be substantial. Averaged over the length of a movie we typically see 20:1 - 60:1 compression for complex, shaded 3D images that fill most of the frame (e.g. Color Plate 26), and 100:1 - 200:1 for simpler 2D images (e.g. Color Plate 25).

8. The Video Animation Server.

The animation server accepts the compressed raster data from the network, decompresses it, and generates and records a video signal. In this server, the software frame buffer is loaded into a hardware frame buffer. The two primary limitations of hardware frame buffers are limited color and spatial resolution.

Limited color resolution limits the subtlety of lighting effects. Color contouring occurs when a surface is supposed to have a continuous graduation in color shade, but the color difference between adjacent pixels needed to produce smooth shading is smaller than the smallest color change that can be represented by the display. The result is bands of color with distinct changes from one band to the next instead of a continuous change. This defect is more noticeable with the hardware frame buffer used because there are only 5 bits per color. The limitation is doubly enforced because the color quantization compression used truncates colors to 5 bits per color [**12**]. Low pass filtering or the addition of high frequency noise would alleviate this problem.

Limited spatial resolution affects the amount of detail that can be represented in an image. The hardware frame buffer resolution is limited (512 columns $\times$ 400 rows pixels) to ensure compatibility with the NTSC standard for video signals.

The video signal generating hardware reads out the image once it is in the hardware frame buffer. In the case of movie making, this video signal is recorded as one of a sequence of video frames on a video recording device. In our implementation the hardware of the animation server consists of an IBM PC/AT™ (or clone), an Ethernet controller used for TCP/IP based IPC, a video frame buffer for reconstructing the raster image and generating the video signal, a single frame animation controller, and a video tape recorder (VTR) (or alternatively a video-optical disk) for recording, and optionally a 68020™ co-processor board for doing the decompression. The hardware configuration is discussed in more detail in [**1**] and [**5**].

There is no user interaction with the PC server. It is used as a peripheral hardware controller serving the system that generates and sends the compressed images. The PC controller is manipulated by means of Sun™ remote procedure calls made from the client. A remote procedure call (RPC) is similar to a conventional procedure call, but is made between processes which are potentially on separate machines. An RPC made by the client causes the invocation of a procedure on the server through the mediation of the RPC package.

RPC calls are made to automatically send an image to the server, decompress it, display it in the hardware frame buffer, and control the video recording device. The procedures on the PC generally operate in parallel with the client: they send an acknowledgement to the client that they are executing, allowing the client to start generating the next image. At the same time the PC is decompressing the image, and displaying it using the local graphics calls. The synchronization necessary, and greater detail on the use of RPC's with this system, are described in [**14**].

Note that this setup is a useful paradigm for capitalizing on the low cost of hardware components used with the PC. Instead of using the PC as a client that makes RPC's to perform computationally intensive tasks, such as to a Cray, the Cray is a client taking advantage of the low-cost peripherals.

In the case where a video-optical disk is used instead of the VTR, the PC also serves as a simple, but useful and convenient, video editing system. The movie clips from an animation sequence are recorded onto the optical disk, and title frames are usually added later, anywhere on the disk. With all of the frames recorded on the disk, a simple script of commands can be loaded into the optical disk unit to control the composition and timing of the final movie. A typical script contains instructions for the video disk to: play the title frame for a certain number of seconds; seek to the graphics clip; hold the first frame for 10 seconds; play the remaining frames at 1/3 speed (10 frames/second animation); and hold the last frame for a few seconds. When the user has constructed a script that produces a satisfactory movie, a VHS recorder is connected to the output of the optical disk, the script is loaded and executed, and the resulting video is recorded to produce a scientific animation that can be viewed on a home VHS player.

9. The Window-Based Server.

Alternatively, the server can run on a window based workstation, such as a Sun™ graphics workstation. On the workstation the graphics is displayed in a window on the screen. There can be more than one window open, so more than one server can run on the workstation. The resolution of such a window can be greater than that of the PC based server. However, if an image is to be shipped to the PC for recording (see below), it will have to be filtered down to the lower resolution.

One problem was encountered because the window-based servers utilize an 8-bit color look-up table. There is no difficulty in decompressing an image when color quantization had been used; color quantization generates the look-up table, and converts each pixel color into an index into the table. However, flickering between frames can occur if the color map changes, which happens frequently when BTC compression is used. BTC compression, while reducing the number of colors in a block of pixels, generates more colors in the image as a whole. This problem can be alleviated somewhat by sorting the color map by luminance during the color quantization process. Since the color map changes, frame-to-frame differencing cannot be used with this window-based server. (The PC server does not use a color map; pixels which do not change are not written over from frame to frame.)

The compressed raster images from the rendering module can be saved on disk by the window-based server. If the images are part of a movie, then when the movie frame generation is complete a preview program can be used to display a portion or the entire sequence on the workstation in forward or reverse at various speeds. The speed of decompressing and displaying a 512 × 400 image is approximately several frames per second on a Sun™ 4-110. The color map problems mentioned above are more noticeable if not corrected, because of the rate of display.

Usually images to be displayed by the previewer are compressed using only BTC and a color map by the client. Lempel–Ziv decompression for each frame will slow the playback speed by more than half. If a movie is not going to be viewed for some period of time, the UNIX compress function, which performs Lempel–Ziv compression, is used on the entire sequence, which has resulted in an observed compression rate on the already compressed images of between 2 and 40 to one.

The previewer has two hardcopy options. In one a specified image from a movie can be converted to a 256 grey level Postscript file and printed on a laser printer. In the other the sequence of compressed images can be sent to the PC server to be decompressed, displayed, and recorded on a video device.

10. The Structure of the User Interface.

The typical user of the movie system is a scientist who needs to produce a movie of the output of a mathematical model. The basic user interface to the distributed movie system is the rendering module. This module implements a low level, GKS like interface for 2D viewing and graphics [**15**], and a SIGGRAPH Core like interface for 3D viewing and graphics [**16**]. To the 3D interface we have added a more intuitive set of 3D viewing controls [**17**], and a polygon primitive with the necessary provisions for shading and lighting. The display of complex 3D objects, such as those illustrated in Color Plate 26, is facilitated by the use of a routine that varies the view position from frame to frame in a way that results in a smooth yaw and pitch rotation. This provides a slow, continuously changing view of the 3D object [**3**].

The distributed aspects of the system are hidden from the user in this interface. Remote procedure calls hide the details of the low-level communications primitives and the server. High-level routines hide the details of RPC's as well. All that the user need know is an identifier for the remote server, and which protocol it is using.

Layered on top of this interface are the graphical representation algorithms, such as Lorensen's Marching Cubes [**6**]. The user interface to the program that uses this algorithm entails nothing more than specification of a 3D grid of function values, and the function value to be displayed (as well as the color, light source position, and initial viewing position). From this specification a complete movie sequence is generated showing the level surfaces undergoing a smooth yaw, pitch and roll rotation to enhance the 3D perception of the surfaces.

The interface to the movie previewing program makes use of the toolkit provided on the window-based server used. A control panel provides a variety of functions. The user clicks on buttons to enable forward or reverse viewing, enters a frame number to seek to, presses a "Postscript" button, and uses a slider to control the speed, among other options.

11. Conclusions.

This system has been successfully used in a number of scientific applications. From the success of this system we conclude: 1) Sophisticated graphics techniques and displays can be made available to the general scientific community by leveraging on the low cost of home video equipment. 2) Distributed computing among remote supercomputers and local workstations is a viable technique even over wide area networks, and can be done transparently to the user. 3) Compression is an essential part of distributed computing.

References

1. W. E. Johnston, D. E. Hall, F. Renema, and D. W. Robertson, *Principles and techniques for low cost computer generated video movies*, Proceedings, Third Computer Graphics Workshop, USENIX Association, Monterey, CA (1986).
2. D. Fink, Editor, "Color Television Standards: Selected Papers and Records of the National Television System Committee," McGraw–Hill, 1955.
3. D. Robertson, *Use of a distributed movie making system for presentation of fluid flow data*, San Francisco State University, San Francisco, CA (Masters Thesis—available as LBL-25274 from Lawrence Berkeley Laboratory) (1988).
4. G. Fischer, Editor, "Mathematical Models," Friedr. Vieweg and Sohn Verlagsgesellschaft mbH, Braumschweig, Germany, 1986.
5. W. E. Johnston, D. E. Hall, J. Huang, M. Rible and D. Robertson, *Distributed scientific video movie making*, Proceedings of the Supercomputing Conference, (The Computer Society of the IEEE) (1988).
6. W. Lorensen and H. Cline, *Marching cubes: A high resolution 3D surface construction algorithm*, Computer Graphics, Vol. 21, No. 4 (Proceedings ACM SIGGRAPH, 1987) (1987).
7. D. Rogers, "Procedural Elements for Computer Graphics," McGraw–Hill, 1985.
8. J. Foley and A. Van Dam, "Fundamentals of Interactive Computer Graphics," Addison–Wesley, 1982.
9. R. Drebin, L. Carpenter and P. Hanrahan, *Volume rendering*, Computer Graphics, Vol. 22, No. 4 (Proceedings ACM SIGGRAPH, 1988) (1988).
10. G. Campbell, T. DeFanti, J. Frederikson, S. Joyce, L. Leske, J. Lindberg and D. Sandin, *Two bit/pixel full color encoding*, Computer Graphics, Vol. 20, No. 4 (Proceedings ACM SIGGRAPH, 1986) (1986).
11. N. Texier, W. Johnston and D. Robertson, *Encoding synthetic animated images*, LBL-24236, University of California, Lawrence Berkeley Laboratory, Berkeley, CA (1987).
12. P. Heckbert, *Color image quantization for frame buffer display*, Computer Graphics, Vol. 16, No. 3 (Proceedings ACM SIGGRAPH, 1982) (1982).
13. T. Welch, *A technique for high performance data compression*, IEEE Computer, Vol. 17, No. 6, June (1984).
14. D. Robertson, W. Johnston, D. Hall and M. Rosenblum, *Video movie making using remote procedure calls and Unix IPC on Unix and UNICOS systems*, LBL-22767, University of California, Lawrence Berkeley Laboratory, Berkeley, CA (1989).
15. G. Enderle, K. Kansy and G. Pfaff, "Computer Graphics Programming: GKS, Second Edition," Springer–Verlag, Berkeley, CA, 1987.
16. Graphics Standards Planning Committee, *Status report of the Graphics Standards Planning Committee*, Computer Graphics, Vol. 13, No. 3 (1979).
17. BJ Wishinsky and W. Johnston, *A simplified interface for SIGGRAPH ore viewing*, LBL-25038, University of California, Lawrence Berkeley Laboratory, Berkeley, CA (1987).

Computing and Information Science Division, Lawrence Berkeley Laboratory, Berkeley, CA 94720 and Department of Mathematics, University of California, Berkeley, CA 94720

Bubbles, Conservation Laws, and Balanced Diagrams

ROB KUSNER

Introduction.

A bubble is one physical system which can be modelled by a constant mean curvature surface in Euclidean three space $\mathbf{R}^3$. Guided by common experience with bubbles, a natural question has been:

> *Must a compact constant mean curvature surface M in $\mathbf{R}^3$ be a round sphere?*

Pioneering research led to a "yes" answer, provided M is either:

CONVEX (*Jellet*)
SIMPLY CONNECTED (*Hopf*)
EMBEDDED (*Alexandrov*)
STABLE (*Barbosa & do Carmo*)

Since these extra hypotheses were thought to be unnecessary, it came as a surprise to many when Wente [**WH**] constructed constant mean curvature tori immersed in $\mathbf{R}^3$.

Recently a general method for constructing complete constant mean curvature surfaces in $\mathbf{R}^3$ was discovered by Kapouleas [**KN**], yielding immersed, compact surfaces of high genus, and also properly embedded non-compact surfaces. At the same time, Korevaar, Kusner, Meeks, and Solomon [**KKS**], [**KKMS**], [**KR**], [**MW**] have developed a structure theory to explain the geometry of these surfaces.

Rather than bubbles, the emerging picture looks more like a model for particle tracks in a bubble chamber!

This note, based upon the author's lectures at MSRI and IMPA in May and August 1988, intends to make the preceding remark precise. It outlines how "elliptic conservation laws" assign "momenta" to a complete constant mean curvature surface in $\mathbf{R}^3$. This leads to a "balanced diagram" along which curves must move to "sweep out" the surface. Thus the balanced structures, around which Kapouleas built examples, actually exist in general.

Symmetry and Conservation.

Suppose M is a surface immersed with constant mean curvature H in a Riemannian 3-manifold $\mathbf{N}$. From the variational viewpoint, M is an extremum for the $Area - H \cdot Volume$ functional. Let $\mathbf{G}$ be the isometry group of $\mathbf{N}$. Variations generated by $\mathbf{G}$ define conserved "momenta" for M as follows. Identify $\mathbf{g}$ (the Lie algebra of $\mathbf{G}$) with 1-parameter groups of isometries of $\mathbf{N}$, and in this way associate to each $Y \in \mathbf{g}$ a unique Killing vector field on $\mathbf{N}$, also denoted Y. Given a 1-cycle Γ on M, pick any 2-chain Δ on $\mathbf{N}$ with $\partial\Delta = \Gamma$. Let ν and η denote appropriately oriented unit normal and conormal vector fields to Δ and Γ, respectively (see figure); we orient the mean curvature vector to be inward-pointing for the sphere. The fundamental result is:

CONSERVATION THEOREM. *Let $\mathbf{N}$ be a Riemannian 3-manifold with trivial first and second homology $H_1(\mathbf{N}) = H_2(\mathbf{N}) = 0$. On any surface M immersed in $\mathbf{N}$ with constant mean curvature H, there is a natural 1-dimensional cohomology "momentum class"*

$$\mu \in H^1(M) \otimes \mathbf{g}^*$$

with coefficients in the (dual) Lie algebra of the symmetry group $\mathbf{G}$ of $\mathbf{N}$, defined by the $Area - H \cdot Volume$ flux

$$\langle \mu([\Gamma]), Y \rangle = \oint_\Gamma \eta \cdot Y - H \iint_\Delta \nu \cdot Y$$

induced by $Y \in \mathbf{g}$ through (Γ, Δ). Moreover, as M is translated in $\mathbf{N}$ by the action of $G \in \mathbf{G}$, the momentum class μ transforms by the coadjoint representation

$$G^*\mu = \mathrm{Ad}^*(G) \cdot \mu.$$

The proof uses the first variation formula, Stokes' Theorem, and the exact homology sequence for a pair. These imply the $Area - H \cdot Volume$ flux through (Γ, Δ) depends solely upon the homology class $[\Gamma] \in H_1(M)$ ("flux is conserved as Γ moves through M") and so μ is well defined in the cohomology $H^1(M) \otimes \mathbf{g}^*$. The transformation property follows from a change of variables in the flux integral.

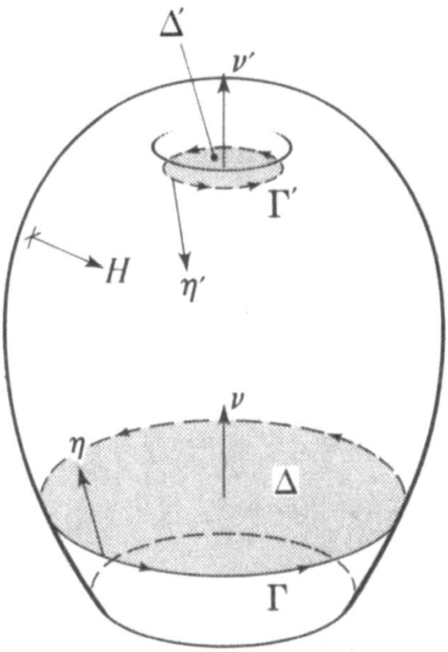

Figure: The $Area - H \cdot Volume$ flux through (Γ, Δ)

Momentum Geometry.

When $\mathbf{N}$ is a spaceform (such as Euclidean $\mathbf{R}^3$, the 3-sphere $\mathbf{S}^3$, or hyperbolic H^3), the momentum class has a useful geometric interpretation. Each coadjoint orbit of $\mathbf{G}$ in $\mathbf{g}^*$ is assigned a pair of real invariants (m, s), and each principal orbit can be identified with the (4-dimensional) space $\mathbf{L}$ of geodesics $L \subset \mathbf{N}$; this gives a map

$$\mathbf{L} \times \mathbf{R}^2 \longrightarrow \mathbf{g}^* : (L, m, s) \longrightarrow \mu$$

which is a diffeomorpism for $m \neq 0$. One can view μ as the (dual of the) infinitesimal motion translating by m along and rotating by s around L. Thus:

MASS–SPIN–WORLDLINE COROLLARY. *Let M be a constant mean curvature surface in a spaceform $\mathbf{N}$. To each 1-cycle $[\Gamma]$ on M, the momentum class μ assigns a "mass" m_Γ, a "spin" s_Γ, and a geodesic "worldline" L_Γ which is unique if $m_\Gamma \neq 0$. If $[\Gamma] = [\Gamma'] + [\Gamma'']$ and the spins vanish, then the* *worldlines L, L' and L'' intersect at a common point $\mathbf{o} \in \mathbf{N}$, and their unit tangent vectors v, v', and v'' at $\mathbf{o}$ "balance" the masses: $mv = m'v' + m''v''$.*

To work this out explicitly for Euclidean space $\mathbf{R}^3$, with symmetry group $\mathbf{E}(3) \approx \mathbf{R}^3 \times \mathbf{O}(3)$, identify the Lie algebra $\mathbf{e}(3)$ and its dual $\mathbf{e}^*(3)$ with

$$\mathbf{R}^3 + \mathbf{o}(3) \approx \mathbf{R}^3 + \mathbf{R}^3,$$

where Lie bracket in the second factor is given by "cross product". The translation and rotation components of $\mu([\Gamma])$ assign "linear momentum" (or "weight")

$$w = \omega([\Gamma]) = \oint_\Gamma \eta - H \iint_\Delta \nu$$

and "angular momentum" (or "torque")

$$r = \rho([\Gamma]) = \oint_\Gamma \eta \times x - H \iint_\Delta \nu \times x$$

vectors to the cycle Γ. (Presumably, these could be measured in a bubble experiment!) The coadjoint action of $(T, R) \in \mathbf{E}(3)$ on $(w, r) \in \mathbf{R}^3 + \mathbf{R}^3 \approx \mathbf{e}^*(3)$ is represented by

$$(T, R)^* w = Rw \quad \text{and} \quad (T, R)^* r = Rr + Rw \times T.$$

For $w \neq 0$, the worldline

$$L_\Gamma = \{T = tw + r \times w / \|w\|^2 \in \mathbf{R}^3 \mid t \in \mathbf{R}\}$$

is the set where $\|(T, 1)^* r\|$ is minimized. If v_Γ is a unit vector orienting L_Γ, the mass and spin invariants are

$$m_\Gamma = w \cdot v_\Gamma \quad \text{and} \quad s_\Gamma = r \cdot v_\Gamma.$$

For example, when M is an embedded Delaunay surface of revolution and Γ is a cross-section, the worldline is the axis of revolution, the spin vanishes, and the mass m_Γ satisfies $0 < m_\Gamma \leq \pi/H$, where the value π/H is attained for the cylinder.

In case $\mathbf{N}$ is the 3-sphere $\mathbf{S}^3$ or hyperbolic space H^3, $\mathbf{G}$ is $\mathbf{O}(4)$ or $\mathbf{O}(3,1)$, and $\mathbf{g}^*$ is identified with the Lie algebra of "antisymmetric" endomorphisms of $\mathbf{R}^4$ or $\mathbf{R}^{3,1}$. The coadjoint action is by conjugation, and so the "eigenvalues" and "eigenspaces" of $\mu([\Gamma])$ determine m_Γ, s_Γ, and L_Γ.

Embedded Surfaces.

If M is embedded in the spaceform $\mathbf{N}$, we can go further. Let $\Pi \subset \mathbf{N}$ be an oriented geodesic plane (open geodesic hemisphere, if $\mathbf{N} = \mathbf{S}^3$) and observe that there is a "positive" Killing vector field Y normal to Π. Let Γ be a union of transverse, compact components of $M \cap \Pi$, and let $\Delta \subset \Pi$ be the planar region bounded by Γ, each inheriting an orientation from Π. If Δ is "outside" M, then the $Area - H \cdot Volume$ flux generated by Y is negative by a direct computation. If Δ is "inside" M, then one can "blow a bubble" B with mean curvature H and with $\partial B = \Gamma$, which is a graph over Δ with respect to the coordinate induced by Y. The flux through (Γ, B) is the same as that through (Γ, Δ). The strong maximum principle applies to show the flux through (Γ, B) is positive, provided $[\Gamma] \neq 0$. (In fact, if $[\Gamma] = 0$, the compact piece of M bounded by Γ coincides with B or its reflection across Π!) It follows that $m_\Gamma \neq 0$, and so there is an oriented worldline L_Γ assigned to nontrivial Γ.

Return now to a properly embedded constant mean curvature $M \subset \mathbf{R}^3$. (Similar results will hold in H^3 and $\mathbf{S}^3$ under extra hypotheses on M.) Using topology it can be shown that the "interior" of M is a "handlebody", that is, a regular neighborhood of a properly embedded 1-dimensional complex W. In many cases, those $[\Gamma]$ dual to an edge of W can be represented as a transverse intersection of M with a plane Π. This leads to a:

BALANCED DIAGRAM. *Let M be a properly embedded constant mean curvature surface in $\mathbf{R}^3$. The mass-spin-worldline collection $\{(m_\Gamma, s_\Gamma, L_\Gamma)\}$ assigned to the 1-cycles $\{[\Gamma]\}$ dual to the edges of W is the "balanced diagram" of M. If the spins all vanish, then segments of the L_Γ can be joined to give a connected graph which balances at each vertex.*

We conjecture that M remains a bounded distance (a universal multiple of $1/H$) from its balanced graph, and (consequently) that the family of all properly embedded constant mean curvature surfaces with a given momentum class μ is compact in the smooth topology. Some progress has been made on this conjecture: the reader is referred to the references for further details, examples and applications.

REFERENCES

[KN] N. Kapouleas, Dissertation, University of California, Berkeley (1988).

[KKS1] N. Korevaar, R. Kusner and B. Solomon, *The structure of complete embedded surfaces with constant mean curvature*, J. Differential Geometry (1989).

[KKS2] ________________, (Work in progress).

[KKMS] N. Korevaar, R. Kusner, W. Meeks and B. Solomon, *Constant mean curvature surfaces in hyperbolic space*, (To appear).

[KR1] R. Kusner, Dissertation, University of California, Berkeley (1988).

[KR2] ________, *Conservation laws for geometric variational problems*, (To appear).

[MW] W. Meeks, *The topology and geometry of properly embedded constant mean curvature surfaces*, J. Differential Geometry (1988).

[WH] H. Wente, *Counterexample to a conjecture of Hopf*, Pacific J. Math. (1986).

Department of Mathematics, Stanford University and the University of California, Santa Barbara

Color Plates

Plates 1–4. Solutions of the generalized Monge–Ampère equation. See "Computer Graphics of Solutions of the Generalized Monge–Ampère Equation" by A. Baldes and O. Wohlrab.

Plate 1. Constant area distortion.

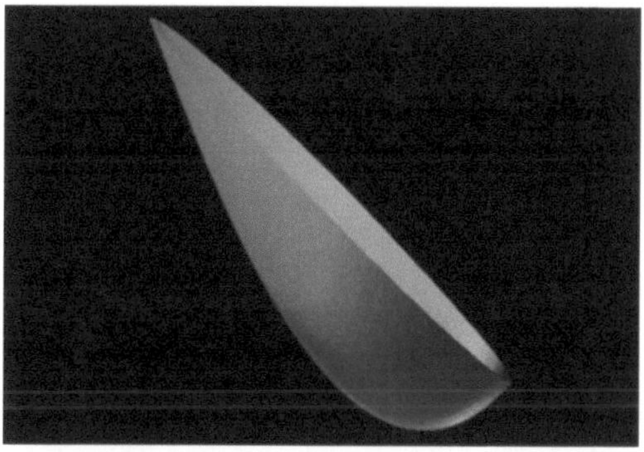

Plate 2. Prescribed spherical area.

Plate 3. Constant Gauss curvature.

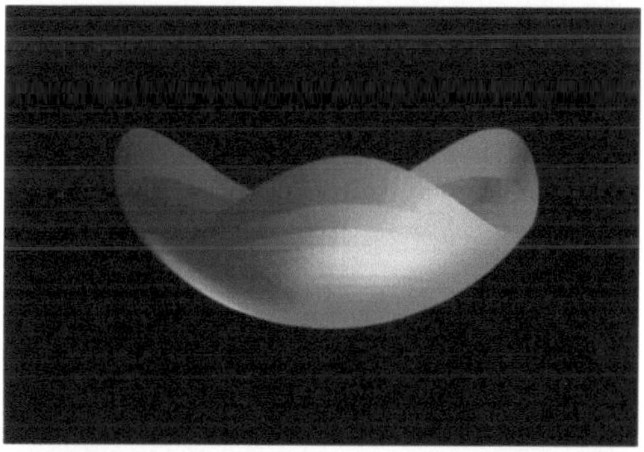

Plate 4. Prescribed Gauss curvature.

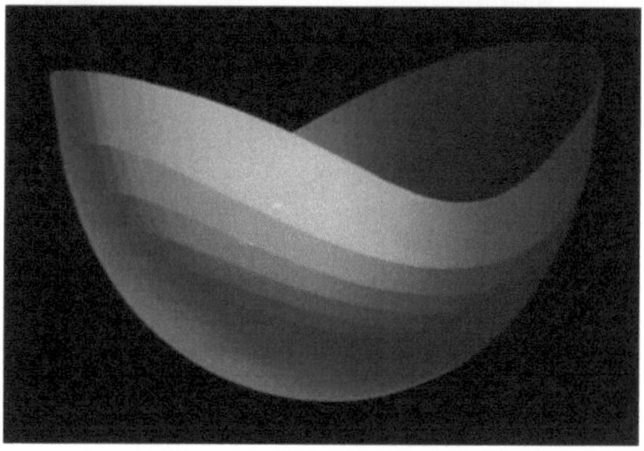

Plates 5–8. Examples of embedded triply-periodic minimal surfaces in the form of plastic models. See "Embedded Triply-Periodic Minimal Surfaces and Related Soap Film Experiments" by A.H. Schoen.

Plate 5. *C(D)*.

Plate 6. *F-RD*.

Plate 7. The gyroid.

Plate 8. *I-WP*.

Plates 9–12. See "Geometric Data for Triply Periodic Minimal Surfaces in Spaces of Constant Curvature" by K. Polthier.

Plate 9. This is a fundamental cell of A.H. Schoen's S′-S″ cell in a cubiod consisting of 16 congruent fundamental pieces. The top and bottom faces of the cubiod are squares. Varying the surface normal at the branchpoint of the Gauss map leads to a 1-parameter family of minimal surfaces in cubiods with different height.

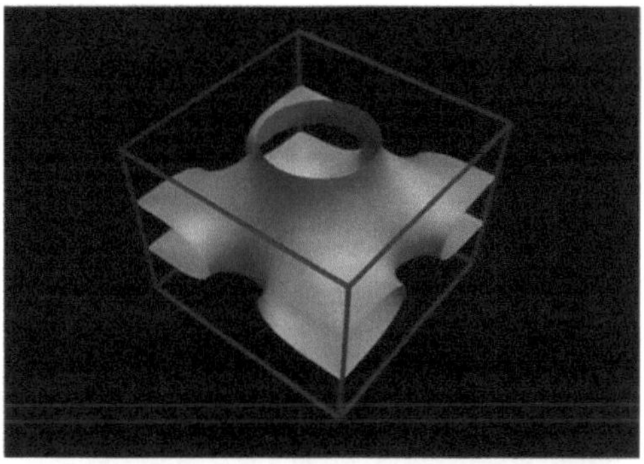

Plate 10. All three space forms, R^3, S^3, and H^3, can be tesselated by cubes. This plate shows four fundamental cells of the classical Schwarz minimal surface. In Plate 11 a cell of a minimal surface in a 72°-cube in hyperbolic 3-space was continued around three edges of the cube [6]. Five cubes are needed to surround each edge.

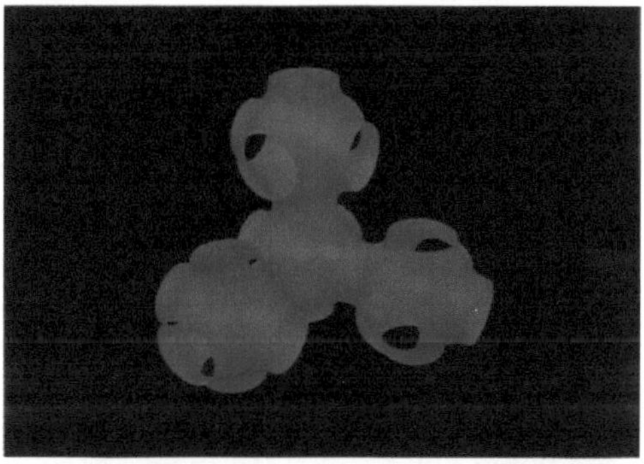

Plate 11. See legend for Plate 10.

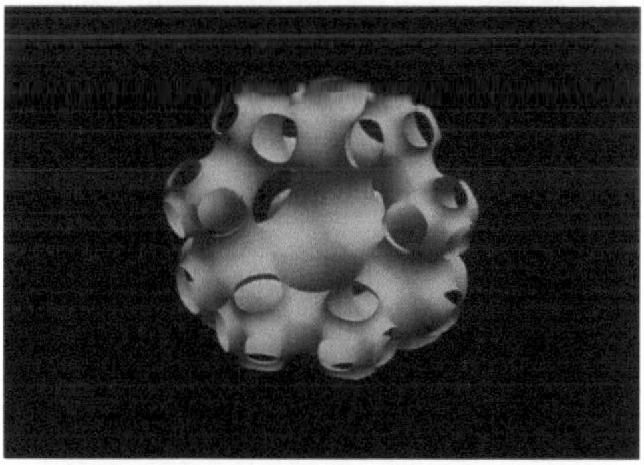

Plate 12. On the *H*-surface of H.A. Schwarz in a hexagonal prism lie straight lines. Three lines connect the upper tips of the horizontal holes and three others connect the lower tips.

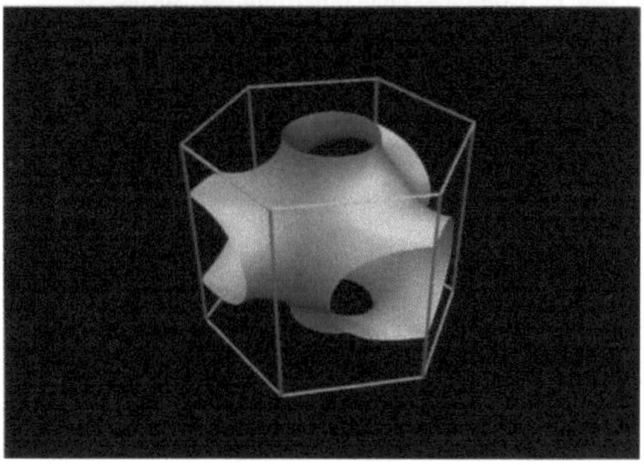

Plates 13–14. See "Geometric Data for Triply Periodic Minimal Surfaces in Spaces of Constant Curvature" by K. Polthier. Plates 15–16. See "From Sketches to Equations to Pictures: Minimal Surfaces and Computer Graphics" by M. Callahan.

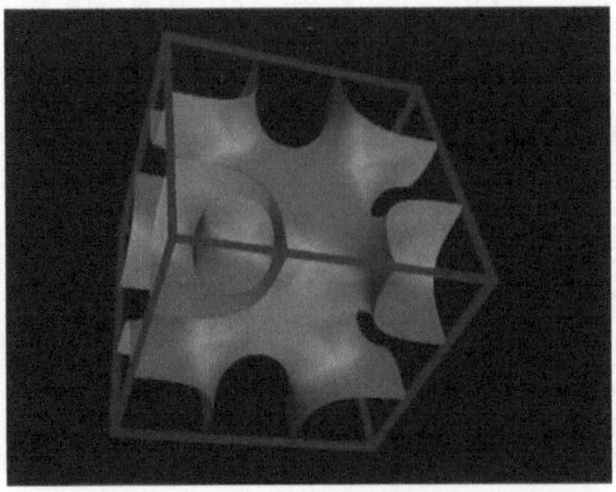

Plate 13. A cell of A. Schoen's *I-WP* surface sits in a cube with a hole to each vertex. The upper half of this cell is a fundamental cell for the group of translations.

Plate 14. The *H'-T* surface divides R^3 into two different labyrinths. One labyrinth is inside the minimal surface. The fundamental cell of the other labyrinth has holes to all faces of a hexagonal prism. It sets on a hexagonal ring in the picture but moved upward half a vertical period.

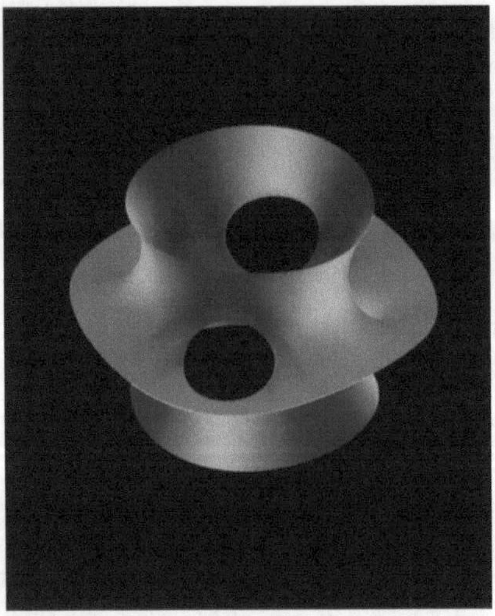

Plate 15. Complete emdedded minimal surface of genus two with three ends.

Plate 16. Complete embedded minimal surface with infinitely many ends.

Plates 17A–18. See "Computer Graphics Tools for Rendering Algebraic Surfaces and for Geometry of Order" by T.F. Banchoff.

Plates 17A–C. The level surfaces of the Čmutov hypersurface in 4-space.

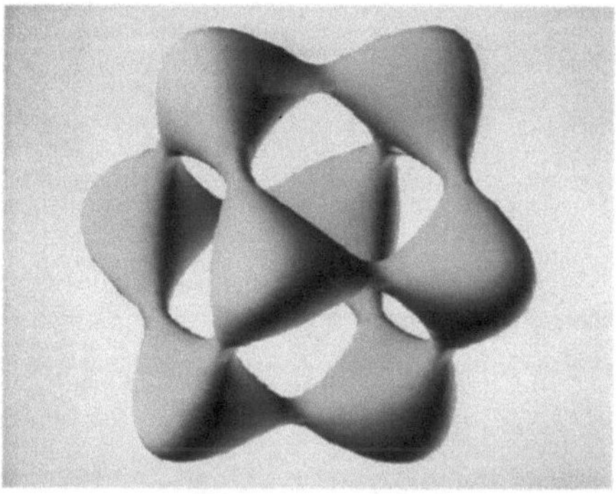

Plate 17B.

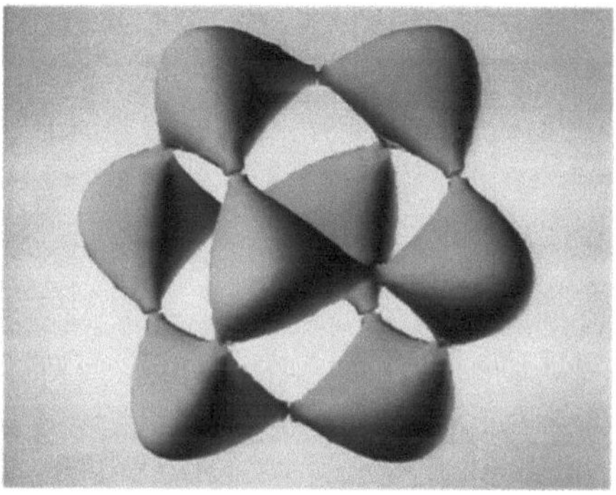

Plate 17C.

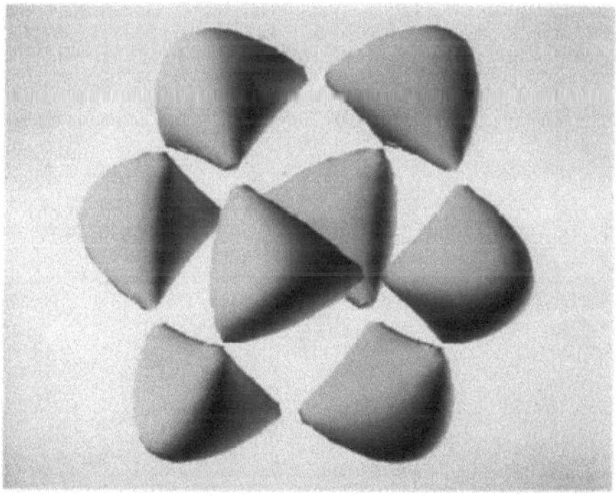

Plate 18. Color-by-order images of a torus of revolution for different directions in space.

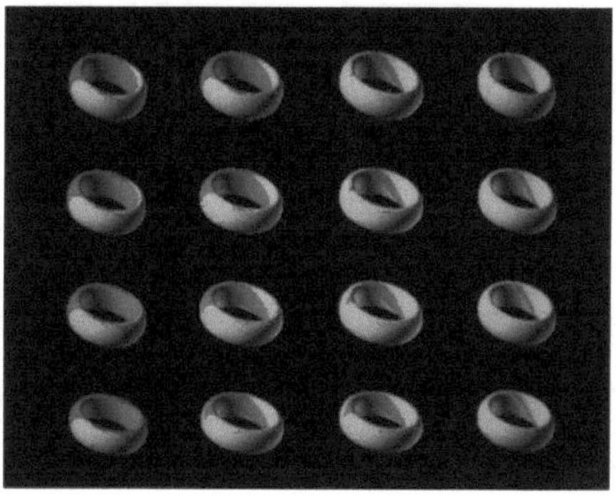

Plates 19–22. See "The Etruscan Venus" by G.K. Francis.

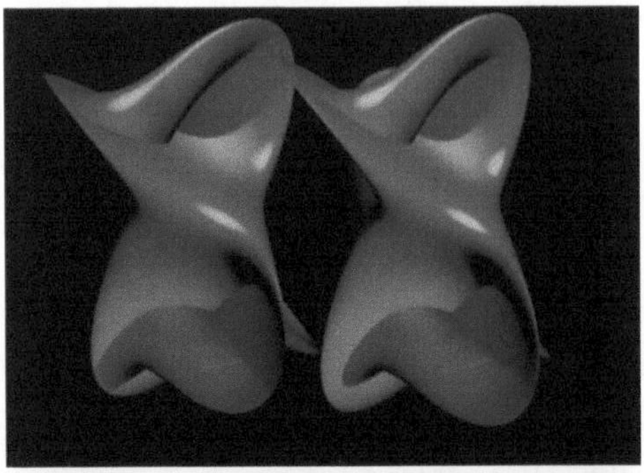

Plate 19. Stable mappings of a Kleinbottle into 3-space just before (left) and after (right) the pinchpoint cancellations.

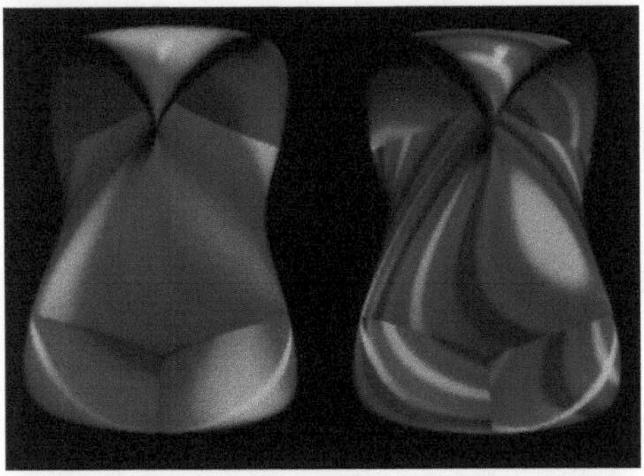

Plate 20. Stable projection of a Kleinbottle embedded in 4-space into 3-space, using a 1-cycle (left) and higher frequency color table to paint the fourth coordinate on the surface (right).

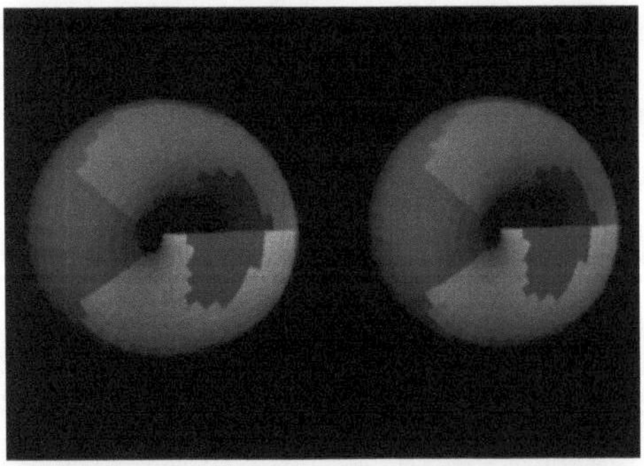

Plate 21. Crossed stereo pair of Steiner's crosscap using five discrete color steps to demonstrate its embedding in 4-space.

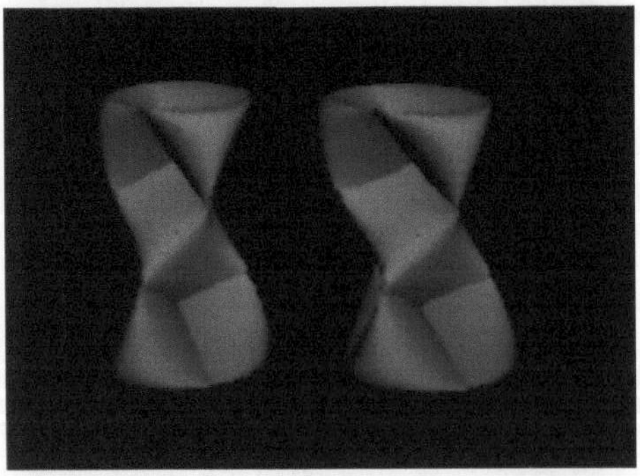

Plate 22. The Etruscan Venus drawn by an experimental 1-chip using a minimal rendering scheme.

Plates 23–24. See "Liquid Menisci in Polyhedral Containers" by D. Langbein and U. Hornung.
Plates 25–26. See "Video Based Scientific Visualization" by W.E. Johnston et al.

Plates 23–24. Capillary surfaces as calculated by the augmented skeleton method.

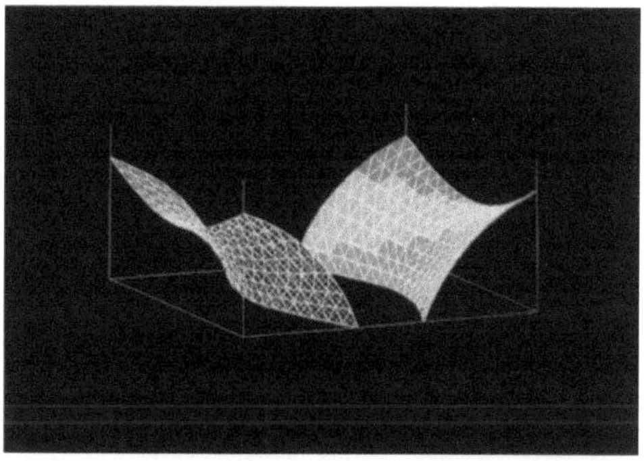

Plate 24.

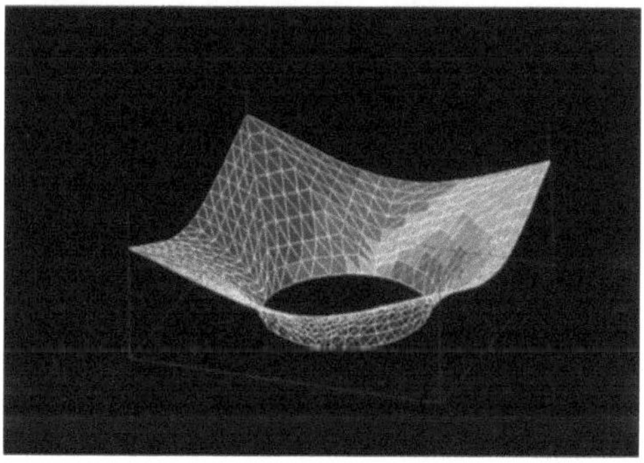

Plate 25. Flow over a backward-facing step.

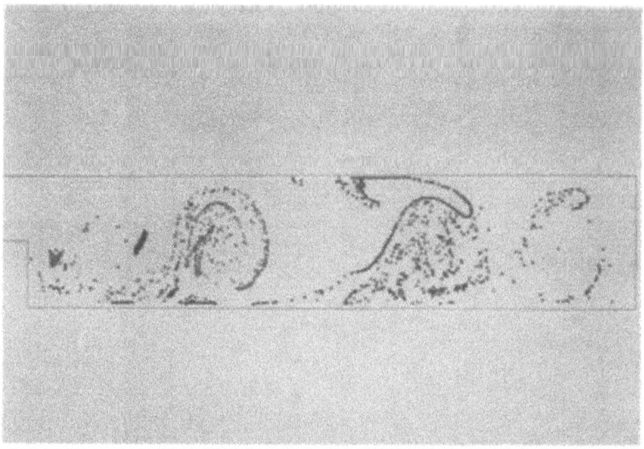

Plate 26. Kummer surface.

Plates 27A–30. Images synthesized using deformable models. See "On Deformable Models" by D. Terzopoulos.

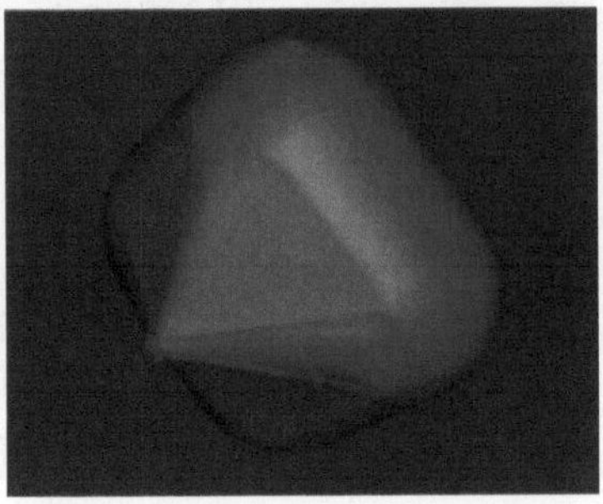

Plates 27A–C. Shrink-wrap. A: Top. B: Middle. C: Bottom.

Plate 28. (Top) A flag waving in the wind.

Plate 29. (Middle) Plasticine bust.

Plate 30. (Bottom) Torn sheet.

Liquid Menisci in Polyhedral Containers

DIETER LANGBEIN AND ULRICH HORNUNG

Abstract. Knowledge of fluid surfaces having constant mean curvature and of their stability limits is a prerequisite for scientific research under microgravity conditions, i.e. in the Spacelab or in the forthcoming Space Station. For the stability of liquid menisci in edges and corners it is essential to know, whether they exhibit negative or positive mean curvature, i.e. whether the meniscus generates an underpressure or an overpressure. A corresponding classification of possible surfaces is attempted. The stability criterion for convex liquid surfaces in long solid edges is indicated and relevant results obtained in parabolic flights of a KC-135 aircraft are reported.

1. Introduction.

Fluid interfaces with constant mean curvature are receiving increasing attention due to the possibilities of living and conducting science and also due to the necessity of solving technical problems in the microgravity environment of space. There is the US Space Shuttle and the USSR Mir Space Station, and the Space Station Freedom is planned for the mid nineties. These space platforms provide microgravity conditions under idealized conditions of zero gravity:

- There is no change in fluidstatic pressure with height.
- There is no buoyancy driven convection in fluids.
- There is neither rising of bubbles nor sedimentation of heavier particles in fluids.

The fuel tank of a spacecraft has to be designed in such a way, that fluid outflow is guaranteed on demand. No bubble is allowed at the fuel outlet, even after several months in orbit. This requirement makes the development and test of surface tension tanks necessary. Equivalently, in scientific or technical experiments with mixtures of metallic melts it is no longer relevant, which melt is the least dense one. The location of melts in a crucible is solely determined by their interface tension and their contact angles with the crucible.

Under conditions of microgravity the shape of liquid menisci may be determined by various methods. Buoyancy has to be small in comparison with the inherent interface tensions, i.e. a low Bond number

$$B = \frac{g\Delta\rho L^2}{\sigma}$$

has to be required. Here, g is the acceleration due to gravity, L is a characteristic length, $\Delta\rho$ the density difference and σ the interface tension between the fluids considered. A low Bond number can be achieved

- by working in a microgravity environment (g small)
- by using density matched liquids (isopygnic liquids, also named Plateau liquids in honor of the blind Belgian physicist Plateau, $\Delta\rho$ small)
- by working with small liquid volumes (L small)
- by performing computer calculations on the shape and stability of liquid menisci
- by analytical solutions of the capillary equation or existence proofs on liquid menisci.

2. Wetting Experiments.

Figures 1 and 2 exhibit the results of two wetting experiments performed under microgravity conditions. They were conducted during parabolic flights of a KC-135 aircraft above the Gulf of Mexico **[1]** **[2]**. Figure 1 shows a rectangular cell with $40mm \times 20mm \times 10mm$ size, which was filled with $\frac{2}{3}$ methanol and $\frac{1}{3}$ cyclohexane. Methanol wets the container walls of aluminum and glass much better than cyclohexane does. Therefore, methanol rapidly spreads around cyclohexane under microgravity conditions. The stopwatch mounted next to the cell shows that spreading of methanol takes less than $5s$. This wetting experiment served as a model experiment on the behaviour of immiscible metallic melts. In sounding rocket experiments on the separation of such melts regularly one component has been found outside, the other one inside.

Figure 2 shows the spreading of fluorinert along the edges of cylinders with rhombic cross-section. The left-hand slimmer cylinder has edges with angles $\frac{\pi}{3}$ and $\frac{2\pi}{3}$, the right-hand cylinder is quadratic, i.e. all edges have angles $\frac{\pi}{2}$. Again, spreading along the edges is very fast. It is faster along the edges with angle $\frac{\pi}{3}$ than along the edges with angle $\frac{\pi}{2}$. Fluorinert reaches the top of the cylinder about $1s$ earlier in the former edges than in the latter edges. It has been proven by Concus and Finn **[3]** **[4]** that a liquid must spread along an edge under microgravity conditions, if half the dihedral angle α of the edge plus the contact angle γ of the liquid to the solid is less than $\frac{\pi}{2}$, i.e., if

$$\alpha + \gamma < \frac{\pi}{2}. \qquad (1)$$

Figure 1: The spreading of methanol around cyclohexane under reduced gravity (during a parabolic flight of a KC-135 aircraft)

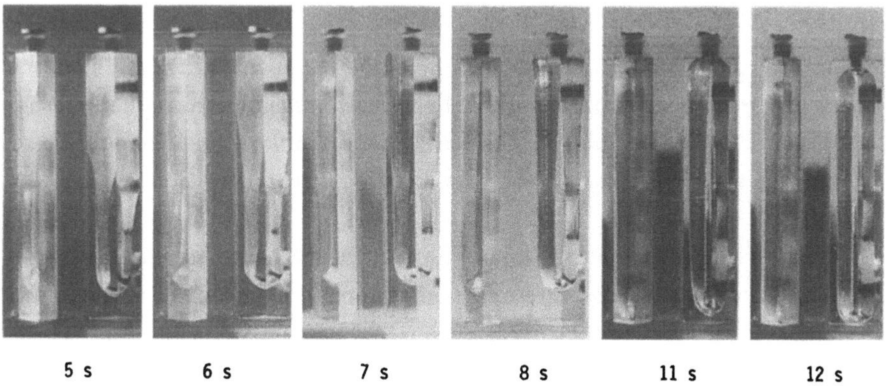

Figure 2: The spreading of fluorinert in cylinders with rhombic cross-section in reduced gravity (during a parabolic flight of a KC-135 aircraft)

Figures 1 and 2 show spreading according to this law to be very fast, even if the respective surface tensions are low. Fluorinert has been selected because of its small contact angle to a large number of solids. Its density is $\rho = 1.8\frac{g}{cm^3}$, its surface tension is much lower than that of water.

3. Menisci with Negative Mean Curvature in Corners.

The basic question regarding the shape of liquid menisci in polyhedral containers in a microgravity environment is, whether there exist corners or edges, in which menisci with negative mean curvature may be formed. Negative curvature in this context refers to looking onto the meniscus from the corner or edge, i.e. a negative mean curvature means that the fluid in the corner or edge has a lower fluidstatic pressure than the fluid in the rest of the container.

The condition for the existence of a meniscus with negative mean curvature located in a rectangular corner reads

$$\cos\gamma > \frac{1}{\sqrt{3}} \text{ or } \gamma < 54.7^\circ,$$

whereas in a rectangular edge one has from Eq. (2)

$$\cos\gamma > \frac{1}{\sqrt{2}} \text{ or } \gamma < 45^\circ.$$

Figure 3a exhibits four menisci with negative mean curvature and contact angle $\frac{1}{\sqrt{2}} > \cos\gamma > \frac{1}{\sqrt{3}}$ in the lower half of a rectangular container. These menisci are Laplace stable, i.e. they satisfy the capillary equation and there is no meniscus deformation directly yielding a configuration with lower energy.

The four liquid volumes shown in Figure 3a nevertheless are—in general—Kelvin unstable. The absolute minimum of energy is assumed, if the liquid volumes in all four corners become equal. The capillary equation, which specifies the state with minimum energy, does not specify the kinetics to reach this minimum. The thermodynamic transition to reach it originates in the vapour pressure. The vapour pressure of a liquid generally increases with increasing curvature of its surface. A small droplet with negative mean curvature therefore causes a larger negative pressure than a large droplet. If several droplets are in a closed container, the large ones tend to decrease the negative pressure, whereas the small ones tend to increase it. The small droplets thus grow on the cost of the large droplets. As a consequence, there is a tendency towards the coexistence of several droplets of equal size.

If liquid volumes in edges with differing dihedral angles are considered, the thermodynamic equilibrium condition reads that the negative mean curvatures (the underpressure) must be equal in all edges.

When the liquid volumes shown in Figure 3a are increased, they will touch each other along the edges. If the container has short edges a, medium edges b and really long vertical edges c, the liquid volumes touch along the short edges first. The menisci shown in Figures 3b and 3c arise. On the one hand, it has to be studied whether these menisci are stable, when the liquid volume is again reduced. On the other hand, by further adding liquid the dry spot on the bottom of Figure 3c will shrink and eventually vanish [5], such that the singly connected liquid volume shown in Figure 3d arises.

4. Menisci with Negative Mean Curvature in Edges.

The described hierarchy of liquid menisci is strongly changed, if equation (1) is satisfied along some or all edges of the container under consideration. The respective edges are being wetted even for very low liquid volumes. The menisci in the corners basically form spherical sections with radius $2R$, those in the edges basically form cylindrical sections with radius R, see Figure 3a. In the vicinity of the corners regions of matching between the spherical and the cylindrical sections arise, i.e. all liquid volumes are connected. They exhibit the same fluidstatic pressure.

When the liquid volume is increased, the liquid volumes in all edges and corners increase uniformly. The dry spot at the small side (a, b) vanishes prior to that at the medium side (a, c). The latter in turn vanishes earlier than that at the large side (b, c). The final menisci reached in the wetting experiments with cyclohexane/methanol (Figure 1) and with fluorinert/air (Figure 2) both touch the small, but do not touch the medium and the large sides. They therefore are indifferent to shifts along the long axis c. In different runs of the wetting experiment with fluorinert/air different volumes of fluorinert have been observed at the small sides. It did not affect the final configuration, whether the experiment was started in horizontal or in vertical orientation of the cylindrical container during the $1.8g$-acceleration phase of the KC-135 aircraft. The time for reaching the final configuration, however, was slightly longer in the horizontal orientation, since the liquid had to spread along the short edges a and b before being able to spread along the long edges c.

5. Menisci with Positive Mean Curvature in Edges.

With increasing contact angle γ the possible menisci adopt a positive mean curvature in the edges and in the corners as well. Several liquid volumes with positive mean curvature in the corners behave like separated drops: They are Laplace stable, but Kelvin unstable. The Kelvin stable situation is a single liquid volume in a single corner.

If the meniscus has positive mean curvature in the corners, it has positive mean curvature in the edges anyway. The Laplace equation enables spherical sections with positive radius $2R$ in the corners and cylindrical sections with positive radius R in the edges and regions of matching in between. Color Plates 23 and 24 exhibit the result of numerical calculations (see the last section of this paper). The pictures were generated on a PC/AT-compatible with an EGA-screen; the colors from red to violet were used instead of shading induced by parallel light.

The question is whether these menisci are stable. They are certainly Kelvin unstable. However, we also have to check on their Laplace stability. Competing configurations are liquid volumes in the corners, in the short edges a, in the medium edges b, etc. It has been shown by Lord Rayleigh already in the last century [**5**] that a free cylindrical jet, the length L of which exceeds the circumference $2\pi R$, is Laplace (or Rayleigh) unstable. There exists a bifurcation of the family of cylindrical surfaces with the family of unduloids. The unduloids have lower energy than the cylinders for $L > 2\pi R$.

An analytical treatment of the bifurcation problem for liquid menisci in solid edges is given in the paper [**6**].

6. Breakage Experiments.

The breakage of cylindrical menisci in solid edges, if $\alpha + \gamma > \frac{\pi}{2}$, has been studied during the KC-135 flights in May and September 1988. Two sets of rhombic cylinders with dihedral angles $(\frac{\pi}{6}, \frac{5\pi}{6})$, $(\frac{\pi}{3}, \frac{2\pi}{3})$ and $(\frac{\pi}{2}, \frac{\pi}{2})$ (see also Figure 2) have been filled with water and with glycerine, respectively.

The cylinders have been used in horizontal orientation, such that during the $1.8g$-acceleration phase of the aircraft the liquid was pressed into the edges with the larger dihedral angles $\frac{5\pi}{6}$, $\frac{2\pi}{3}$ and $\frac{\pi}{2}$, respectively. In reduced gravity the liquid not only moved to the cylinder's ends, but in the edges with dihedral angle $\frac{5\pi}{6}$ and $\frac{2\pi}{3}$ also left single droplets in the middle. This breakage could be reproduced in several parabolae. In the edge with dihedral angle $\frac{\pi}{2}$ no breakage was observed.

Figure 4 shows the respective breakage in circular cylinders in two different parabolae. Plexiglass cylinders with $4mm$, $6mm$, $7mm$, $9.8mm$, and $12.5mm$ in diameter (from top to bottom) and $90mm$ in length have been filled with $\frac{1}{5}$ coloured water. The upper cylinder leaked by accident. In the second and third cylinder the water was not pressed to the bottom even during the $1.8g$ deceleration and acceleration phases. It just formed tongues along the bottom, which rapidly withdrew during microgravity. The single droplets visible in these cylinders result from earlier handling. In the two lower cylinders water floated along the bottom during $1.8g$ and $1g$ as expected. Breakage in microgravity required about $5s$ and $15s$ in the cylinders with diameter $9.5mm$ and $12.5mm$, respectively. The wavelength of breakage obviously was shorter in the cylinder exhibiting the larger diameter.

The exact stability criteria for convex liquid menisci in short edges, either separated or joined, still have to be evaluated. It is expected that this can be achieved by numerical methods.

7. Description of the Numerical Method.

The numerical technique used to obtain Figure 3 and Color Plates 23 and 24 is the augmented skeleton method ASM reports on which are going to be published in [7] [8]. Minimizers of the energy functional

$$E = \int_\Gamma \sigma d\Gamma - \int_\Sigma \cos\gamma d\Sigma$$

with the volume constraint

$$V = \text{const}$$

are calculated. Here, Γ is the capillary surface and Σ the wetted part of the container walls.

The basic idea of ASM is to discretize the unknown surface using finite elements and to find the stationary points of the resulting finite dimensional optimization problem. Γ is dealt with as a perturbation of a skeleton which is made up by a finite number of triangles. The directions of perturbation are prescribed vectors in space. Thus, both the energy E and the volume V become functionals of finitely many parameters. Once the variational problem has been discretized, the numerical problem is to find minimizers, or, at least, stationary points. To this end, the differentials of E and V are calculated analytically. Then, the system of equations

$$\partial E + p\partial V = 0, \; V = \text{ const}$$

is solved for the mentioned parameters and p, where the pressure p is the Lagrange multiplier. This is achieved by using the classical Newton method, which was chosen for three reasons:

- the Jacobian of the system can be calculated relatively easily, since the functionals involved are given explicitly,
- the Jacobian is symmetric; hence special methods can be applied, which take advantage of symmetry and sparseness,
- any possible sources of convergence problems have to be avoided, since the radius of convergence may be very small.

As a standard tool for nonlinear problems a continuation method, often called a homotopy method, is used to find initial guesses for Newton's method. One starts with a situation for which the exact solution of the variational problem is explicitly known. Naturally, these are configurations with spheres or parts of spheres as capillary surfaces. Then, one changes the parameters involved, such as the angles or the volume in small steps, and in this way solves a whole sequence of problems. The strategy is to use the solution of the old problem as a starting guess for the new one. Of course, in order to speed up the procedure more sophisticated extrapolation techniques can be used.

The method ASM easily accounts for gravity and/or rotation. No convergence problems were encountered unless in cases where one is extremely close to nonexistence of the capillary surface.

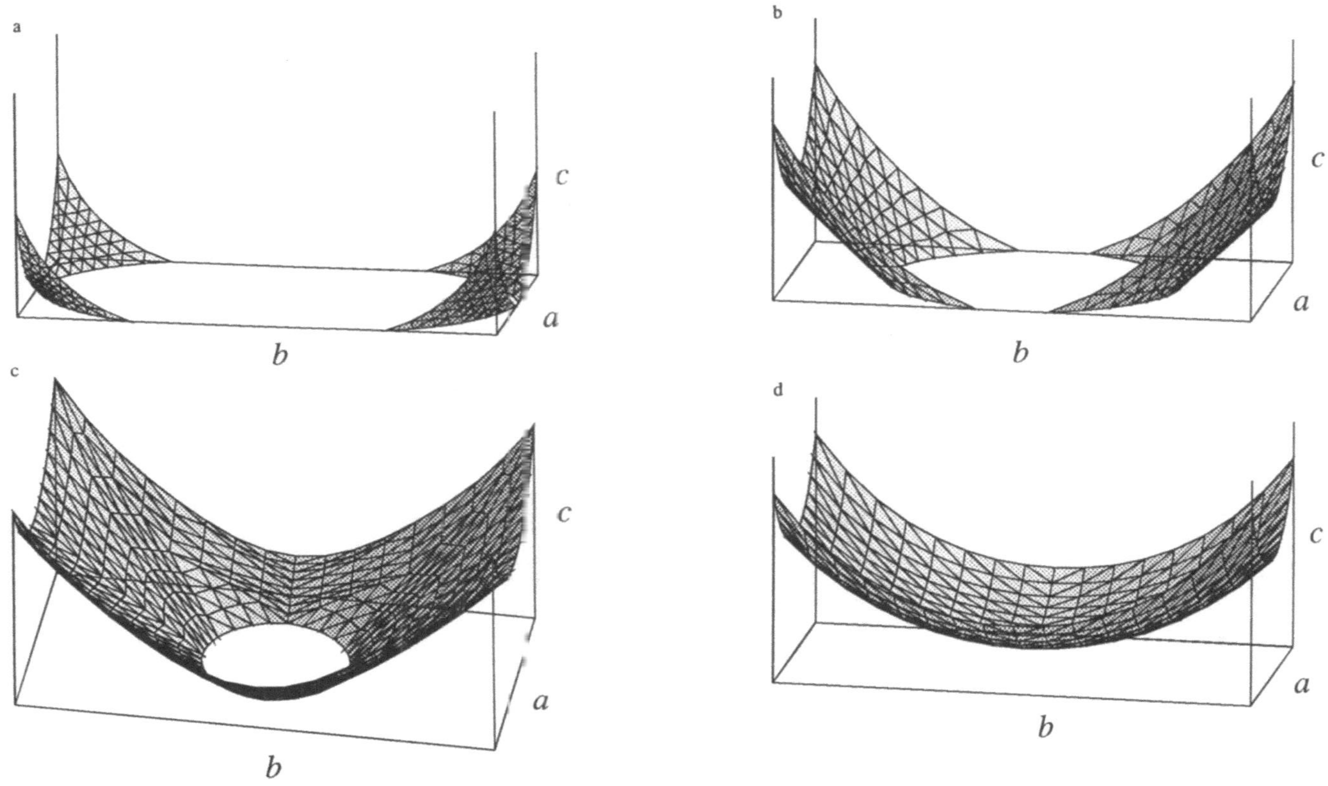

Figure 3: Liquid menisci with negative mean curvature at the bottom of a rectangular container. If the separated volumes shown in (a, b) have differing radii, there is evaporation from the large to the small volumes. When the liquid volume is increased, the initial volumes in the corners touch and join along the short edges a and subsequently along the medium edges b. Eventually the dry spot in the centre vanishes.

Acknowledgements.

We gratefully acknowledge sponsoring of preparation, performance and evaluation of the parabolic flight experiments by the German Bundesministerium für Forschung und Technologie under contract number QV 8723. The wetting experiments were conceived in cooperation with R. Finn, Stanford and W. Heide, Frankfurt. R. Großbach, Frankfurt and H. Schlegel, G. Thiele, H. Walpot, Köln alternately assisted the performance.

References

1. Langbein, D. and Heide, W., *Fluid physics demonstration experiments*, in "Science Demonstration Experiments during Parabolic Flights of KC-135 Aircraft", ESTEC-WP 1457 (1986), 47–54.
2. ______, *Study of convective mechanisms under microgravity conditions*, Adv. Space Res. **6/5** (1986), 5–17.
3. Concus, P. and Finn, R., *On capillary free surfaces in the absence of gravity*, Acta Math. **132** (1974), 177–198.
4. Finn, R., *Equilibrium capillary surfaces*, Grundlehren der Mathematischen Wissenschaften **284** Springer (1986), 1–244.
5. Rayleigh, J. W. S., Lord, *On the capillary phenomena of jets*, Proc. Roy. Soc. **29** (1879), 71–97.
6. Langbein, D., *The shape and stability of liquid menisci in solid edges*, J. Fluid Mech. (submitted).
7. Hornung, U. and Mittelmann, H. D., *A finite element method for capillary surfaces with volume constraints*, J. Comput. Phys. **87** (1990), 126–136.
8. ______, *The augmented skeleton method for parametrized surfaces of liquid drops*, J. Colloid Interface Sci. **133** (1989), 409–417.

Dieter Langbein, Battelle Institut, Am Römerhof 35, D-6000 Frankfurt 90, Fed. Rep. Germany
Ulrich Hornung, SCHI, P.O. Box 1222, D-8014 Neubiberg, Fed. Rep. Germany

Can One Hear the Shape of a Fractal Drum? Partial Resolution of the Weyl–Berry Conjecture

MICHEL L. LAPIDUS*

1. Introduction.

Several years ago, motivated in part by the challenging problem of studying the scattering of light from "fractal" surfaces, the physicist Michael V. Berry formulated a very intriguing conjecture about the vibrations of "drums with fractal boundary". Extending to the "fractal" case a long standing conjecture of Hermann Weyl, he conjectured in particular that the high frequencies of such "fractal drums" were governed by the Hausdorff dimension of their boundary [**1,2**].

In this paper, we describe a recent work [**13**] in which we have obtained a partial resolution of the Weyl–Berry conjecture.

Our techniques derive from methods of partial differential equations, the calculus of variations and — to a lesser extent — geometric measure theory. An interesting aspect of this work is that it lays the foundations for a theory connecting spectral geometry and "fractal" geometry.

Although we consider both Dirichlet and Neumann boundary conditions as well as positive elliptic operators of order $2m$, we first present our results for the more familiar case of the Dirichlet Laplacian, for the simplicity of exposition. Let Ω be a bounded open set of $\mathbf{R}^n$ ($n \geq 1$) with "fractal" boundary Γ. We consider the following eigenvalue problem:

$$\text{(P)} \qquad \begin{cases} -\Delta u = \lambda u \text{ in } \Omega, \\ \quad u = 0 \text{ on } \Gamma, \end{cases}$$

where $\Delta = \sum_{k=1}^{n} \partial^2/\partial x_k^2$ denotes the Laplace operator. Here, the Dirichlet problem (P) is understood in the *variational sense*; that is, we say that the scalar λ is an eigenvalue of (P) if there exists $u \neq 0$ in $H_0^1(\Omega)$ [the closure of $C_0^\infty(\Omega)$, the space of C^∞ functions with compact support in Ω, in the Sobolev space $H^1(\Omega)$] such that $-\Delta u = \lambda u$, in the distributional sense. [Recall that $H^1(\Omega)$ denotes the set of functions u in $L^2(\Omega)$ with

*Research partially supported by the National Science Foundation under Grant DMS-8703138.

(distributional) gradient ∇u also in $L^2(\Omega)$, equipped with the Hilbert norm $\|u\|_{H^1(\Omega)} := \left(\|u\|^2_{L^2(\Omega)} + \|\nabla u\|^2_{L^2(\Omega)}\right)^{1/2}$.]

It is well known that the spectrum of (P) is discrete and consists of an infinite sequence of eigenvalues:

$$0 < \lambda_1 \leq \lambda_2 \leq \cdots \leq \lambda_j \leq \ldots, \text{ with } \lambda_j \to +\infty \text{ as } j \to \infty.$$

Physically, these eigenvalues and their associated eigenfunctions represent, for example, the natural frequencies and steady-states vibrations of a "drum" Ω with boundary Γ.

The asymptotic behavior of the large eigenvalues can be deduced from that of the "counting function":

$$N(\lambda) = \text{card}\{j \geq 1 : \lambda_j \leq \lambda\},$$

the number of eigenvalues (counted with multiplicity) of (P) not exceeding $\lambda > 0$.

H. Weyl [**25**] has proved that if Ω is a bounded open set in $\mathbf{R}^n$, with sufficiently smooth boundary, then

$$N(\lambda) \sim (2\pi)^{-n}\mathcal{B}_n|\Omega|_n\lambda^{\frac{n}{2}}, \text{ as } \lambda \to +\infty, \tag{1.1}$$

where $\mathcal{B}_n$ denotes the volume of the unit ball in $\mathbf{R}^n$ and $|\Omega|_n$ the n-dimensional Lebesgue measure or "volume" of Ω. (See also, e.g., [**5,20,23**].)

In his classical paper, entitled "Can one hear the shape of a drum?," M. Kac [**12**] posed the following question: can someone with perfect pitch recognize the shape of a drum just by listening to the fundamental tone and all the overtones?

This question gave rise to many beautiful results connecting, in particular, differential geometry and analysis (see, e.g., [**17**]).

If the boundary Γ is smooth (i.e., of class C^∞), Seeley [**24**]—and then, when $n \geq 4$, Pham The Lai [**22**]—has shown that (1.1) could be supplemented by the following remainder estimate:

$$N(\lambda) = (2\pi)^{-n}\mathcal{B}_n|\Omega|_n\lambda^{\frac{n}{2}} + 0(\lambda^{\frac{n-1}{2}}), \text{ as } \lambda \to +\infty; \tag{1.2}$$

this result—based on the theory of pseudodifferential operators—extended, in particular, earlier work of Hörmander [**9**]. [Here and thereafter, the notation "$f(\lambda) = O(g(\lambda))$, as $\lambda \to +\infty$" means that there exist positive constants λ_0 and C such that $|f(\lambda)| \leq Cg(\lambda)$, for all $\lambda \geq \lambda_0$.]

If Γ is C^∞ [and if, in addition, $\bar{\Omega}$ does not have too many multiply reflected closed geodesics], Ivrii [**11**] has proved that Weyl's 1912 conjecture holds true (see also [**19**] and [**10**]):

$$(1.3)\quad N(\lambda) = (2\pi)^{-n}\mathcal{B}_n|\Omega|_n\lambda^{\frac{n}{2}} - c_n|\Gamma|_{n-1}\lambda^{\frac{n-1}{2}} + o(\lambda^{\frac{n-1}{2}}), \text{ as } \lambda \to +\infty;$$

here, c_n denotes a positive constant and $|\Gamma|_{n-1}$ the $(n-1)$-dimensional volume (e.g., the length if $n = 2$ or the area if $n = 3$) of Γ.

Berry [**1,2**]—motivated in part by the study of the scattering of waves from "fractal" or random surfaces and the study of porous media—extended Weyl's conjecture to the case when the boundary Γ is "fractal" as follows:

$$(1.4)\quad N(\lambda) = (2\pi)^{-n}\mathcal{B}_n|\Omega|_n\lambda^{\frac{n}{2}} - c_{n,H}\mathcal{H}_H(\Gamma)\lambda^{\frac{H}{2}} + o(\lambda^{\frac{H}{2}}), \text{ as } \lambda \to +\infty;$$

here, the real number H in $[n-1, n]$ denotes the Hausdorff dimension Γ and $\mathcal{H}_H(\Gamma)$ the H-dimensional Hausdorff measure of Γ.

Recently, Brossard and Carmona [**3**] found a simple counterexample to (1.4) and suggested that the Hausdorff dimension H should be replaced by the less familiar Minkowski dimension D. [That D is a better suited notion of "fractal" dimension is also clear retrospectively, in view of [**8**] in conjunction with [**13**, esp. Proposition 3.1] and is explained in detail in [**13**].]

The Weyl-Berry conjecture has numerous physical applications, including, for example [**1,2**], to the study of the vibrations of a "fractal drum" or of water in a lake [$n = 2$ and $D \in (1,2)$], as well as the oscillations of the Earth or the acoustic modes of a concert hall with very irregular walls [$n = 3$ and $D \in (2,3)$].

2. Fractal Dimensions.

We now briefly recall the definition and the main properties of the Minkowski dimension. (See [**13**, §2.1 and §3], as well as the references therein for more precise information.)

DEFINITION 2.1. *For $\epsilon > 0$, let Γ_ϵ, the (open) ϵ-neighborhood of Γ, be the set of $x \in \mathbf{R}^n$ within a distance $< \epsilon$ from Γ. Let $D = D(\Gamma)$ be the infimum of the positive numbers d such that*

$$(2.1)\qquad \mathcal{M}_d = \mathcal{M}_d(\Gamma) := \limsup_{\epsilon \to 0+} \epsilon^{-(n-d)}|\Gamma_\epsilon|_n < +\infty.$$

D (resp., $\mathcal{M}_D$) is called the Minkowski dimension (resp., content) of Γ.

REMARKS 2.1: (a) We have $\mathcal{M}_d = +\infty$ for $d < D$ and $\mathcal{M}_d = 0$ for $d > D$. Moreover, we have in general $\mathcal{M}_D \in [0, +\infty]$. [For most of the classical "fractals" considered in [**4,6,18**], however, we have $0 < \mathcal{M}_D < +\infty$.]

(b) The larger D, the more irregular Γ. Further, we always have $D \in [n-1, n]$ since $\Gamma = \partial\Omega$ (see [**13**, Corollary 3.2]). *We shall say here that Γ is "fractal" if $D \in (n-1, n]$*; this is the case in particular if D is not an integer.

(c) When the boundary Γ is sufficiently "regular" (e.g., $(n-1)$-rectifiable [**7**] and, in particular, of class C^1), we have $H = D = n-1$, the topological dimension of Γ; in general, however, $n-1 \leq H \leq D \leq n$ (e.g., [**13**, Proposition 3.2]). [Another case when $H = D$ is when Γ is (strictly) "self-similar", in the sense of [**18**] (see, e.g., [**13**, Lemma 3.1]); this fact seems to have been the source of much confusion among the practitioners of fractal geometry. A simple example for which H and D differ strikingly is provided in [**13**, Example 5.1].

(d) Intuitively, the Hausdorff (resp., Minkowski) dimension can be understood as follows: let $\mathcal{N}(\epsilon)$ be the number of n-dimensional cubes of diameter $\leq \epsilon$ (resp., $= \epsilon$) needed to cover Γ; then, very roughly, if $\mathcal{N}(\epsilon)$ increases like $\mathcal{N}(\epsilon) \propto \epsilon^{-H}$ (resp., ϵ^{-D}), as $\epsilon \to 0^+$, Γ has Hausdorff (resp., Minkowski) dimension H (resp., D). (Compare [**13**, Definition 3.1] and [**13**, Corollary 3.1] for a precise version of this statement.) This basic difference between H and D will enable us to probe the inner structure of Γ in this problem.

3. Partial Resolution of the Weyl–Berry Conjecture.

We now present, in a special case, the main result established in [**13**]. (See [**13**, Theorem 2.1 and Corollary 2.1].) It constitutes a significant step towards a resolution of the Weyl–Berry conjecture.

THEOREM 3.1. *Let Ω be an arbitrary (nonempty) bounded open set in $\mathbf{R}^n$, with boundary Γ.*

(i) *If Γ is "fractal" (i.e., if $D \in (n-1, n]$), then for any $d > D$,*

$$N(\lambda) = (2\pi)^{-n}\mathcal{B}_n|\Omega|_n\lambda^{\frac{n}{2}} + 0(\lambda^{\frac{d}{2}}), \text{ as } \lambda \to +\infty. \tag{3.1}$$

(ii) *If Γ is "nonfractal" (i.e., if $D = n - 1$), then for any $d > D$,*

$$(3.2) \qquad N(\lambda) = (2\pi)^{-n}\mathcal{B}_n|\Omega|_n\lambda^{\frac{n}{2}} + 0(\lambda^{\frac{d}{2}}\log\lambda), \text{ as } \lambda \to +\infty.$$

Moreover, except possibly in the degenerate case when $\mathcal{M}_D(\Gamma) = +\infty$, the remainder estimate (3.1) [resp., (3.2)] also holds in case (i) [resp., (ii)] with $d = D$, the Minkowski dimension of Γ.

REMARKS 3.1: (a) We stress that we do not make here about Γ any hypothesis of self-similarity (or, more generally, self-alikeness), in the sense of Mandelbrot [**18**].

(b) The "nonfractal" case when $\mathcal{M}_D(\Gamma) < +\infty$ was obtained earlier by Métivier in [**20**], using a different terminology, and under much more restrictive assumptions, by Courant (see [**5**, p. 445]). Prior to our results, (3.2) [with $d = D = n - 1$] was the best estimate known for "nonsmooth" bounded open sets. Note, of course, that a "nonfractal" boundary Γ may be far from smooth.

(c) As will be recalled below, we also show in [**13**, Examples 5.1–5.1′] that our estimates in the "fractal" case are best possible.

According to Definition 2.1 and the comments following it, Theorem 3.1 is deduced from the next result, also established in [**13**]:

THEOREM 3.2. *Let $d \in (n-1, n]$ (resp., $d = n - 1$) be such that $\mathcal{M}_d(\Gamma) < +\infty$. Then (3.1) [resp., (3.2)] holds for this value of d.*

In [**13**], we also obtain the counterpart of Theorems 3.1 and 3.2 for general positive elliptic operators of order $2m$ ($m \geq 1$) on Ω : $\mathcal{A} = \sum_{|\alpha|\leq m, |\beta|\leq m}(-1)^{|\alpha|}D^\alpha(a_{\alpha\beta}(x)D^\beta)$, with $a_{\alpha\beta} = \overline{a_{\beta\alpha}}$ and with (locally) constant leading coefficients $a_{\alpha\beta}$ ($|\alpha| = |\beta| = m$). (See [**13**, Theorem 2.1] for a more precise statement.) [Naturally, in the above discussion, we have set $\mathcal{A} = -\Delta$; the special case of the Dirichlet Laplacian was announced in [**15**].]

Moreover, we also consider the rather delicate case of Neumann (or, more generally, mixed Dirichlet–Neumann) boundary conditions. Hence, under appropriate assumptions that guarantee that the Neumann spectrum is discrete and that its leading asymptotics obey Weyl's law (1.1), we obtain the analogue of Theorems 3.1 and 3.2. Such hypotheses are of two kinds:

(a) Either that Ω satisfies the so-called "(C') condition"; that is, roughly speaking, if its boundary Γ is not "too long". This is the case, for example, if Ω obeys a "segment property" (e.g., if Γ is Lipschitz) or else if Ω is a bounded open set with cusp. (For case (a), see [**13**, Theorem 2.1 and Corollary 2.2].)

(b) Or that Ω has the "extension property"; that is (when $m = 1$), if there exists a bounded linear extension operator from $H^1(\Omega)$ to all of $H^1(\mathbf{R}^n)$. This is the case, for example, of "quasidisks" when $n = 2$ or of their higher-dimensional analogues ("Jones domains"), arising in the study of conformal geometry and harmonic analysis. It is noteworthy that the boundary Γ can then have any Hausdorff dimension in $[n-1, n)$. (For case (b), see [**13**, Theorem 4.1].)

We also deduce from Theorem 2.1 corresponding remainder estimates for the short time asymptotics of the "partition function" or trace of the heat semigroup [**13**, Theorem 2.2].

Our proof of Theorem 3.2—given in [**13**, §4]—is purely analytic. Partly motivated by [**8**], it extends that of [**5**] and [**20**] to the "fractal" case. Variational methods—based in particular on the max-min formula for the ith eigenvalue, the method of "Dirichlet–Neumann bracketing", as well as the Kolmogorov i-width from approximation theory—play a crucial role. We cover Ω by small cubes whose size decreases to zero as one approaches the boundary. This enables us, in some sense, to connect at a "micro-local" level the spectral and "fractal" geometries of $\bar{\Omega}$. A very delicate step consists in obtaining precise boundary estimates in terms of the Minkowski dimension D of Γ.

In the course of the proof, we introduce naturally the notion of "*conjugate fractional exponents*" of the boundary Γ: namely, $\theta := D - (n-1)$ and $\theta' := 1 - \theta = n - D$, which both lie in $[0, 1]$. The surprising dichotomy between the "fractal" and "nonfractal" cases observed in the statement of Theorems 3.1 and 3.2 can then be attributed technically to the fact that the partial sums of a geometric series of ratio θ take a different form according to whether $\theta \neq 0$ or $\theta = 0$. We shall propose a deeper interpretation of this dichotomy in a later work.

In [**13**], we show by means of examples that our estimates are optimal in every possible "fractal" dimension; more precisely, we construct a one-parameter family of examples for which our remainder estimates are sharp and the Minkowski dimension D of Γ takes on every value in $(n-1, n)$. (See [**13**, Examples 5.1–5.1′]; Berry's original conjecture obviously fails for these examples.) We also formulate a modification of the Weyl–Berry conjecture. (See [**13**, §5.2].)

In a work in preparation [**16**], Carl Pomerance and the author prove this "modified Weyl–Berry conjecture" in the *one-dimensional* case ($n = 1$) and illustrate it, in particular, for the aforementioned family of examples studied in [**13**]. This leads us to establish some unexpected and intriguing connections between our previous work [**13**] and aspects of analytic number theory, particularly the theory of the Riemann zeta–function.

4. Final Remarks.

We conclude this paper by several suggestions for the direction of future research. We have considered in this work the case of a "drum with fractal boundary". In this situation, methods from the calculus of variations as well as techniques going beyond ordinary (smooth) differential geometry played a major role. Actually, Berry's original conjecture [**1,2**] was also formulated in the mathematically ill-defined case of a "fractal drum" (Ω is itself "fractal" and hence no longer open). We hope to tackle this problem in a later work [**14**]. A related problem is to analyze mathematically wave phenomena on "fractals": for example, to study the scattering of light from "fractal" surfaces. We expect that a successful investigation of this challenging problem will require new insight gained from the use of computer graphics. In turn, it is likely that the area of computer graphics would benefit from such a study.

REFERENCES

1. M. V. Berry, *Distribution of modes in fractal resonators*, in "Structural Stability in Physics", W. Güttinger and H. Eikemeir (eds.), Springer–Verlag, Berlin (1979), 51–53.
2. ———, *Some geometric aspects of wave motion: wavefront dislocations, diffraction catastrophes, diffractals*, in "Geometry of the Laplace Operator", Proc. Symp. Pure Math., Vol. 36, Amer. Math. Soc., Providence, RI (1980), 13–38.
3. J. Brossard and R. Carmona, *Can one hear the dimension of a fractal?*, Comm. Math. Phys. **104** (1986), 103–122.
4. G. Cherbit (ed.), "Fractals, Dimensions Non Entières et Applications," Masson, Paris, 1987.
5. R. Courant and D. Hilbert, "Methods of Mathematical Physics," Vol. I, Interscience, New York, NY, 1953.
6. H. Falconer, "The Geometry of Fractal Sets," Cambridge Univ. Press, Cambridge, 1985.
7. H. Federer, "Geometric Measure Theory," Springer, Berlin, 1969.
8. J. Fleckinger and M. L. Lapidus, *Remainder estimates for the asymptotics of elliptic eigenvalue problems with indefinite weights*, Arch. Rational Mech. Anal. **98** (1987), 329–356.

9. L. Hörmander, *The spectral function of an elliptic operator*, Acta Math. **121** (1968), 193–218.
10. ———, "The Analysis of Linear Partial Differential Operators," Vols. III and IV, Springer–Verlag, Berlin, 1985.
11. V. Ja. Ivrii, *Second term of the spectral asymptotic expansion of the Laplace-Beltrami operator on manifolds with boundary*, Functional Anal. Appl. **14** (1980), 98–106.
12. M. Kac, *Can one hear the shape of a drum?*, Amer. Math. Monthly **73** (1986), 1–23.
13. M. L. Lapidus, *Fractal drum, inverse spectral problems for elliptic operators and a partial resolution of the Weyl–Berry conjecture*, Univ. of Georgia preprint, Athens (1988), 123 pages; to appear in the *"Transactions of the American Mathematical Society"*.
14. ———, *Elliptic differential operators on fractals and the Weyl–Berry conjecture*, in preparation.
15. M. L. Lapidus and J. Fleckinger–Pellé, *Tambour fractal: vers une résolution de la conjecture de Weyl–Berry pour les valeurs propres du laplacien*, C. R. Acad. Sci. Paris Sér. I Math. **306** (1988), 171–175.
16. M. L. Lapidus and C. Pomerance, *The Riemann zeta-function and the one-dimensional Weyl-Berry conjecture for fractal drums*, in preparation. [To be announced in "M. L. Lapidus and C. Pomerance, *Fonction zêta de Riemann et conjecture de Weyl–Berry pour les tambours fractals*, C. R. Acad. Sci. Paris Sér. I Math." (to appear)].
17. H. P. McKean and I. M. Singer, *Curvature and the eigenvalues of the Laplacian*, J. Differential Geom. **1** (1967), 43–69.
18. B. B. Mandelbrot, "The Fractal Geometry of Nature," rev. and enl. ed., W. H. Freeman, New York, NY, 1983.
19. R. B. Melrose, *Weyl's conjecture for manifolds with concave boundary*, in "Geometry of the Laplace Operator", Proc. Symp. Pure Math., Vol. 36, Amer. Math. Soc., Providence, RI (1980), 254–274.
20. G. Métivier, *Valeurs propres de problèmes aux limites elliptiques irréguliers*, Bull. Soc. Math. France Mém. **51–52** (1977), 125–219.
21. H. O. Peitgen and P. H. Richter, "The Beauty of Fractals," Springer–Verlag, Berlin, 1986.
22. Pham The Lai, *Meilleures estimations asymptotiques des restes de la fonction spectrale et des valeurs propres relatifs au laplacien*, Math. Scand. **48** (1981), 5–38.
23. M. Reed and B. Simon, "Methods of Modern Mathematical Physics," Vol. IV, Academic Press, New York, NY, 1978.
24. R. T. Seeley, *A sharp asymptotic remainder estimate for the eigenvalues of the Laplacian in a domain of* $\mathbb{R}^3$, Adv. in Math. **29** (1978), 244–269.
25. H. Weyl, *Das aysmptotische Verteilungsgesetz der Eigenwerte linearer partieller Differentialgleichungen*, Math. Ann. **71** (1912), 441–479.

Department of Mathematics, The University of Georgia, Boyd Graduate Studies Research Center, Athens, GA 30602 U.S.A.

Steepest Descent as a Tool to Find Critical Points of $\int k^2$ Defined on Curves in the Plane with Arbitrary Types of Boundary Conditions

ANDERS LINNÉR

Abstract. We will explicitly compute the gradient of the total squared curvature functional on a space of parametrized curves of fixed or variable length satisfying arbitrary types of boundary conditions. We show how to turn the space of such curves into an infinite dimensional submanifold of an inner product space. The steepest descent will then be along the integral curves of the negative gradient vector field in this manifold. We will derive the gradients and the corresponding flow equations. In conclusion we use computer graphics to illustrate this process by following one such trajectory starting close to an unstable critical point and ending at a stable critical point.

0. Introduction.

This paper will develop an infinite dimensional method for finding critical points of the total squared curvature functional $\int k^2$. We will restrict our attention to parametrized curves in the plane sufficiently smooth so the above integral exists. The following three types of boundary conditions will be considered:

I. Both end points fixed;

II. I and the direction at one of the end points given;

III. I and the directions at both end points given.

We will also introduce a factor representing the length of the curves. If the variation in this factor is set to zero it corresponds to fixing the length of the curves.

The following two special cases have been studied before:

Boundary conditions III and variable length under the assumption that the curves are graphs of functions. This rules out curves with self intersections as well as other examples (see [**1**]).

Boundary conditions III and fixed length assuming closed curves, i.e., the end points are the same and the directions coincide there (see [**2**], [**3**] and [**4**]).

Some of the interest in the total squared curvature functional stems from the fact that among the critical points are the geodesics, i.e., curves with curvature zero everywhere. Of course in the plane the only geodesics are the straight lines or line segments. However being part of a straight line may be incompatible with the boundary conditions so we expect to find other critical points, the elastic curves. By holding the end points of a piece of a springy wire in different configurations we can produce examples of such elastic curves. The purpose of this paper is to derive equations, suitable for implementation, whose solutions can be used to find the elastic curves.

Given a system of n equations $f_s(x) = 0$, $s \in \{1, \ldots, n\}$, $x \in R^n$ we note that $F(x) = \sum_{s=1}^{n} f_s^2(x)$ has a global minimum zero at x if and only if x is a solution to the above system. A way of attacking the above system is to choose an arbitrary x and then take a step in the direction given by $-\nabla F(x)$. We will generalize this steepest descent in two ways. First, our x will be an element of an infinite dimensional inner product space representing planar curves γ. We replace f_s by $k(s)$ the curvature at $\gamma(s)$, here s is the arc length parameter. We want to follow the integral curves (trajectories) of $-\nabla F(\gamma)$ where $F(\gamma) = \int_\gamma k^2(s)ds$. Note that $\nabla F(\gamma)$ will be an element of the inner product space. Secondly, the above boundary conditions will determine subsets of the inner product space that turn out to be infinite dimensional submanifolds. In general $\nabla F(\gamma)$ is not expected to be in the tangent space to this submanifold so we will show how to take the tangential part.

In conclusion it should be pointed out that it is not automatic that a given sequence γ_n of curves on which F is bounded and where $\nabla F(\gamma_n)$ converges to zero has a convergent subsequence. For the submanifolds considered in this paper this property has been established in [**2**].

The paper is organized as follows:

0. Introduction
1. Spaces and functionals
2. Gradients
3. Flow equations and an example

The author wishes to thank Wojbor Woyczynski, David A. Singer, Joel Langer and the rest of the Math department at CWRU for creating an opportunity to work on this paper.

1. Spaces and Functionals.

We will use the following conventions throughout the paper. Let $I = [0, 1]$

be the parameter interval, $\gamma : I \longrightarrow R^2$ a curve in the plane. We sometimes write $\gamma = (x, y)$ and we let $(a, b) = \gamma(1) - \gamma(0)$ be the vector between the end points. Letting $'$ denote differentiation with respect to the parameter we assume $|\gamma'| = L$. Here $|\ |$ is the standard Euclidean norm and L is the length of the curve. We consider both fixed and variable L but the above means that γ is parametrized proportional to arc length so L is fixed for each curve but may vary along the trajectory.

Next we introduce the indicatrix of γ which is the unique continuous function $\theta : I \longrightarrow R$ satisfying $\theta(0) \in [0, 2\pi)$ as well as $\gamma' = L(\cos\theta, \sin\theta)$. The signed curvature $k : I \longrightarrow R$ is now given by $k = \frac{1}{L}\theta'$. Let $L^2(I)$ denote the space of real valued functions on I whose square is Lebesgue integrable and let

$$H = \{\theta : I \longrightarrow R \mid \theta \text{ absolutely continuous, } \theta' \in L^2(I)\}.$$

Absolute continuity implies differentiability almost everywhere (see [**5**] pages 145–148). We want $\theta' \in L^2(I)$ in order for our functional to make sense. Note that H includes indicatrices representing curves whose curvature has a finite number of jump discontinuities.

H will represent curves of fixed length $L = 1$. To handle the variable length case we define:

$$\widehat{H} = H \times \{L \in R \mid L > \sqrt{a^2 + b^2}\}.$$

For each λ a positive real number we define the following functional on $\widehat{H}$:

$$\widehat{J}_\lambda = \frac{1}{2L}\int_I \theta'^2 + \lambda L.$$

This functional corresponds to the total squared curvature functional. The $\frac{1}{2}$ will simplify the derivative formulas and we interpret λL as a penalty for long curves. The three types of boundary conditions are now characterized by:

I. $L\int_I \cos\theta = a$ and $L\int_I \sin\theta = b$;
II. I and $\theta(0) = \theta_0$, where $\theta_0 \in [0, 2\pi)$;
III. II and $\theta(1) = \theta_1$, where $\theta_1 \in R$.

The following inner products will be used on H and $\widehat{H}$:

$$\langle\theta, \widetilde{\theta}\rangle_H = \int_I \theta'\widetilde{\theta}' + \theta(0)\widetilde{\theta}(0) \quad \langle(\theta, L), (\widetilde{\theta}, \widetilde{L})\rangle_{\widehat{H}} = \langle\theta, \widetilde{\theta}\rangle_H + L\widetilde{L}.$$

With the metric induced by $\langle\ ,\ \rangle_H$, H is a complete space, i.e., a Hilbert space.

It turns out that the subsets of $\widehat{H}$ corresponding to the above boundary conditions are closed submanifolds. To understand why this is the case we define two closed subspaces of H:

$$H_0 = \{\theta \in H \mid \theta(0) = 0\}, \qquad H_{00} = \{\theta \in H_0 \mid \theta(1) = 0\}.$$

Also let

$$\begin{aligned} H_{0t} &= \{\theta \in H \mid \theta(0) = \theta_0,\ \theta_0 \in [0, 2\pi)\}, \\ H_{00t} &= \{\theta \in H_{0t} \mid \theta(1) = \theta_1,\ \theta_1 \in R\} \end{aligned}$$

be linear translates of these subspaces so they are in fact closed submanifolds of H. As we did for H we introduce a factor for variable length L and we get $\widehat{H}_{0t}$ and $\widehat{H}_{00t}$. Let $\Phi : \widehat{H} \longrightarrow R^2$ be given by $\Phi(\theta, L) = (L \int \cos\theta, L \int \sin\theta)$. We want to show that the derivative of Φ restricted to $\widehat{H}_{00t}$ is surjective (onto) on $\Phi^{-1}(a, b)$. This also shows that when Φ is restricted to $\widehat{H}_{0t}$ or not restricted at all the derivative is surjective. Applying a consequence of the implicit function theorem (see **[6]** page 162) this will show that the subsets corresponding to the different boundary conditions are closed submanifolds. Another consequence is that the tangent space will split into a two dimensional factor and the kernel of the derivative of Φ, both closed subspaces. This fact will be used later when finding the tangential part of the gradient. Proceeding to show the surjectivity we begin with

LEMMA 1.1. *If h and k are two continuous linearly independent real valued functions defined on I then there exist two continuous real valued functions f and g defined on I such that $f(0) = f(1) = g(0) = g(1) = 0$ and $\int_I fh \int_I gk \neq \int_I fk \int_I gh$.*

PROOF:

$$\int_I (h-k)h \int_I (h+k)k - \int_I (h+k)h \int_I (h-k)k =$$
$$= 2(\int_I h^2 \int_I k^2$$

Here the inequality is a consequence of the linear independence and Cauchy–Schwarz inequality.

By connecting $(0,0)$ to $(\frac{1}{n},(h+k)(\frac{1}{n}))$ and $(1-\frac{1}{n},(h+k)(1-\frac{1}{n}))$ to $(1,0)$ for $n = 3,4\ldots$ using straight line segments we generate a sequence of continuous functions g_n. Similarly for $h-k$ we get a sequence f_n. Since g_n differ from $h+k$ and f_n from $h-k$ on sets of measure no more than $\frac{2}{n}$ and this difference is decreasing uniformly, it follows that for sufficiently large n

$$\int_I f_n h \int_I g_n k > \int_I g_n h \int_I f_n k,$$

which proves the lemma.

To conclude the submanifold argument we assume that $\Phi(\theta, L) = (a,b)$ then the derivative of Φ in the direction $(v_\theta, v_L) \in H \times R$ is:

$$D\Phi(\theta, L)(v_\theta, v_L) = (\frac{v_L a}{L} - L\int_I v_\theta \sin\theta, L\int_I v_\theta \cos\theta).$$

To show that $D\Phi(\theta, L)$ is surjective it suffices to exhibit two linearly independent vectors $(\int_I v_\theta \sin\theta, \int_I v_\theta \cos\theta)$ and $(\int_I \widetilde{v}_\theta \sin\theta, \int_I \widetilde{v}_\theta \cos\theta)$. But $\sin\theta$ and $\cos\theta$ are linearly independent because else θ would be constant which is impossible if $L > \sqrt{a^2+b^2}$, (the curve is not a straight line segment). An application of Lemma 1.1 and its proof concludes the argument because if $v_\theta \in H$ vanishes at the end points then v_θ is in the tangent space to H_{00t}.

We now give an alternative way of characterizing the tangent spaces. First we define another set of functionals on $\widehat{H}$. Let λ_1 and λ_2 be real numbers and let

$$\widehat{\Lambda}(\theta, L) = L\int_I(\lambda_1 \cos\theta + \lambda_2 \sin\theta).$$

If $\Phi(\theta, L) = (a,b)$ then $D\Phi(\theta, L)(v_\theta, v_L) = 0$ if and only if $D\widehat{\Lambda}(\theta, L)(v_\theta, v_L) = 0$ for all λ_1 and λ_2. As we will see in the next section this statement implies that the gradient of $\widehat{\Lambda}(\theta, L)$ is perpendicular to the tangent space at (θ, L) for all λ_1 and λ_2. This will be the key to finding the tangential part of the gradient of $\widehat{J}_\lambda$.

2. Gradients.

Analogous to the finite dimensional case we define the gradient of a real valued function F defined on an inner product space by the relation

$$\langle \nabla F(x), v\rangle = DF(x)v.$$

Here $\langle\ ,\ \rangle$ denotes the inner product and x, v are arbitrary vectors. Using the notation established in Section 1 we introduce components by $(\alpha_\theta, \alpha_L) = \nabla \widehat{J}_\lambda(\theta, L)$ and $(\beta_\theta, \beta_L) = \nabla \widehat{\Lambda}(\theta, L)$. Remembering that $(x', y') = L(\cos\theta, \sin\theta)$ we get

PROPOSITION 2.1. *For the boundary cases* I, II *and* III *the second components of the gradients are given by*

$$\alpha_L = \lambda - \frac{1}{2L^2}\int_0^1 (\theta'(s))^2 ds$$

$$\beta_L = \frac{1}{L}(\lambda_1 a + \lambda_2 b)$$

and with

$$A(s) = \int_0^s (\lambda_1(y(t) - y(0)) - \lambda_2(x(t) - x(0)))dt,$$

the first components of the gradients for the different boundary cases are

$$\begin{aligned}
\text{I.} \quad & \alpha_\theta(s) = \tfrac{1}{L}(\theta(s) - \theta(0)) \\
& \beta_\theta(s) = A(s) + (\lambda_1 a - \lambda_2 b)(s+1) \\
\text{II.} \quad & \alpha_\theta(s) = \tfrac{1}{L}(\theta(s) - \theta_0) \\
& \beta_\theta(s) = A(s) + (\lambda_1 a - \lambda_2 b)s \\
\text{III.} \quad & \alpha_\theta(s) = \tfrac{1}{L}(\theta(s) - (\theta_1 - \theta_0)s - \theta_0) \\
& \beta_\theta(s) = A(s) - A(1)s.
\end{aligned}$$

Before proving the proposition we remark that the above formulas are invariant under translation but case III simplifies considerably if the origin is chosen at the centre of gravity of the curve, i.e., one assumes $\int_I \gamma = 0$ (this fact was explored in [4]). It also turns out that if in case II we assume $\theta(1) = \theta_1$ instead of $\theta(0) = \theta_0$ the formulas are more complicated. This is due to the choice of inner product which in our case has a term involving evaluation at 0. We now recall the duBois–Reymond lemma: If $f : I \longrightarrow R$ is continous and $\int_0^1 f(s)v'(s)ds = 0$ for all $v : I \longrightarrow R$ such that v' is continous and $v(0) = v(1) = 0$ then f is constant. The proof consists of the clever observation that $v(s) = \int_0^s (f(t) - \int_0^1 f(u)du)dt$ satisfies the above requirements. Since $v'(s) = f(s) - \int_0^1 f(u)du$ we get $0 = \int_0^1 (v'(s) + \int_0^1 f(u)du)v'(s)ds = \int_0^1 (v'(s))^2 ds$ the clever part being to substitute for f rather than v'.

PROOF (2.1): The derivatives at $(\theta, L) \in \widehat{H}$ in the direction $(v_\theta, v_L) \in H \times R$ are given by

$$D\widehat{J}_\lambda(\theta,L)(v_\theta,v_L) = \frac{1}{L}\int_0^1 \theta'(s)v_\theta'(s)ds + (\lambda - \frac{1}{2L^2}\int_0^1 (\theta'(s))^2 ds)v_L,$$

$$D\widehat{\Lambda}(\theta,L)(v_\theta,v_L) = \int_0^1 (\lambda_1(y(s)-y(0)) - \lambda_2(x(s)-x(0)))v_\theta'(s)ds$$

$$+ (\lambda_2 a - \lambda_1 b)v_\theta(1) + \frac{1}{L}(\lambda_1 a + \lambda_2 b)v_L.$$

If we look at directions $(v_\theta, 0) \in H \times R$ and use the inner product $\langle\ ,\ \rangle_{\widehat{H}}$ then the equalities defining the gradients become

$$\int_0^1 \alpha_\theta'(s)v_\theta'(s)ds + \alpha_\theta(0)v_\theta(0) = \frac{1}{L}\int_0^1 \theta'(s)v_\theta'(s)ds,$$

$$\int_0^1 \beta_\theta'(s)v_\theta'(s)ds + \beta_\theta(0)v_\theta(0) = \int_0^1 (\lambda_1(y(s)-y(0)) - \lambda_2(x(s)-x(0)))v_\theta'(s)ds$$

$$+ (\lambda_2 a - \lambda_1 b)v_\theta(1).$$

If we moreover assume v_θ' continuous and $v_\theta(0) = v_\theta(1) = 0$ then an application of the duBois–Reymond lemma and a subsequent integration shows that

$$\alpha_\theta(s) = \frac{1}{L}\theta(s) + cs + \widetilde{c}$$

$$\beta_\theta(s) = A(s) + ds + \widetilde{d}$$

for some constants $c, \widetilde{c}$ and $d, \widetilde{d}$. To find the correct constants we again use the equations defining the gradients and get

$$c(v_\theta(1) - v_\theta(0)) + (\frac{1}{L}\theta(0) + \widetilde{c})v_\theta(0) = 0$$

$$d(v_\theta(1) - v_\theta(0)) + \widetilde{d}v_\theta(0) - (\lambda_2 a - \lambda_1 b)v_\theta(1) = 0.$$

By choosing v_θ consistent with the various boundary conditions we finally conclude for

I,II $v_\theta(0) = 0,\ v_\theta(1) \neq 0$ implies $c = 0$ and $d = (\lambda_2 a - \lambda_1 b)$,
I $v_\theta(0) \neq 0$ then implies $\widetilde{c} = -\frac{1}{L}\theta(0)$ and $\widetilde{d} = d$,
II,III $\alpha_\theta(0) = 0,\ \beta_\theta(0) = 0$ implies $\widetilde{c} = -\frac{1}{L}\theta_0$ and $\widetilde{d} = 0$,
III $\alpha_\theta(1) = 0,\ \beta_\theta(1) = 0$ implies $c = \frac{1}{L}(\theta_1 - \theta_0)$ and $d = -A(1)$.

Also note that the direction $(0,1) \in H \times R$ determines α_L and β_L which concludes the proof.

We now turn our attention to how the tangential part of $\nabla \widehat{J}_\lambda$ is determined. The facts given at the end of Section 1 show that β_θ is perpendicular to the tangent space of the submanifold corresponding to one of the boundary conditions. It was also noted in Section 1 that the space of all vectors perpendicular to the tangent space is two dimensional. If $(\lambda_1, \lambda_2) \mapsto \beta_\theta$ is viewed as a map from the plane into this two dimensional space it is actually linear. To see that it is surjective we observe that for case I, $(1,0) \mapsto \int_0^s (y(t)-y(0))dt - b(s+1)$ and $(0,1) \mapsto -\int_0^s (x(t)-x(0))dt + a(s+1)$. Now if one of these functions is a scalar multiple of the other so is its derivative. This however implies that $x(s)$ and $y(s)$ satisfy a linear relation which is a contradiction since none of our curves are straight lines. The same argument works for the other boundary cases as well. All in all there is a unique (λ_1, λ_2) such that $\nabla \widehat{J}_\lambda - \nabla \widehat{\Lambda}$ is tangent, in particular it is in the kernel of $D\Phi$ which gives the following vector equation

$$\int_0^1 \beta_\theta(s)\gamma'(s)ds + \frac{\beta_L}{L}(b,-a) = \int_0^1 \alpha_\theta(s)\gamma'(s)ds + \frac{\alpha_L}{L}(b,-a).$$

Here the (λ_1, λ_2) dependent terms are collected on the left hand side, and the equation is linear in (λ_1, λ_2). We will not attempt to solve explicitly for (λ_1, λ_2) but assume this is done to get the flow equations in the next section.

3. Flow Equations and an Example.

Steepest descent in our context will be to follow the integral curve (trajectory) of $\nabla \widehat{\Lambda} - \nabla \widehat{J}_\lambda$. This trajectory will be a curve in the submanifold determined by the boundary conditions. We let w be the parameter along the trajectory and we use it as subscript to indicate dependence along the trajectory. This convention together with the correct choice of (λ_1, λ_2) according to the end of Section 2 leads to the following flow equation for the indicatrix

$$\frac{\partial \theta_w}{\partial w}(s) = \nabla \widehat{\Lambda}_w(\theta_w, L_w)(s) - \nabla \widehat{J}_\lambda(\theta_w, L_w)(s).$$

There is little hope, except in very special cases (for examples see [**4**]), to solve this equation explicitly. The good news however is that the equation is sufficiently well behaved that elementary numerical methods give accurate solutions. To illustrate this we offer what we feel is a particularly spectacular example. It is known (see [**2**]) that for closed curves with indicatrices satisfying $\theta(0) = \theta(1)$ there is only one stable critical point, a certain figure eight. Moreover the only other critical points are multiple covers of this figure eight (using different scales). Our example (see Figures 3.1 and 3.2) consists of different curves along a trajectory starting at a closed curve with reflectional but not rotational symmetry close to the double covered critical figure eight. Note that the scale is changed in Figure 3.2 and the example lacks the symmetry needed to be solved explicitly.

The flow equation was solved using a combination of Eulers method and two types of integration. The fundamental numerical problem when finding integral curves is that only a finite number of points can be used to describe the curves along the trajectory. We elected to space our points according to Gauss–Legendre between parameters where the curvature of the initial curve was discontinuous. The trapezoid rule was then used for the first level of integrals and Gauss–Legendre quadrature for subsequent levels. The advantage with Euler's method is that it automatically gives a sense of the dynamics along the trajectory. This has been the basis for some animated computer graphics developed by the author.

Figure 3.1

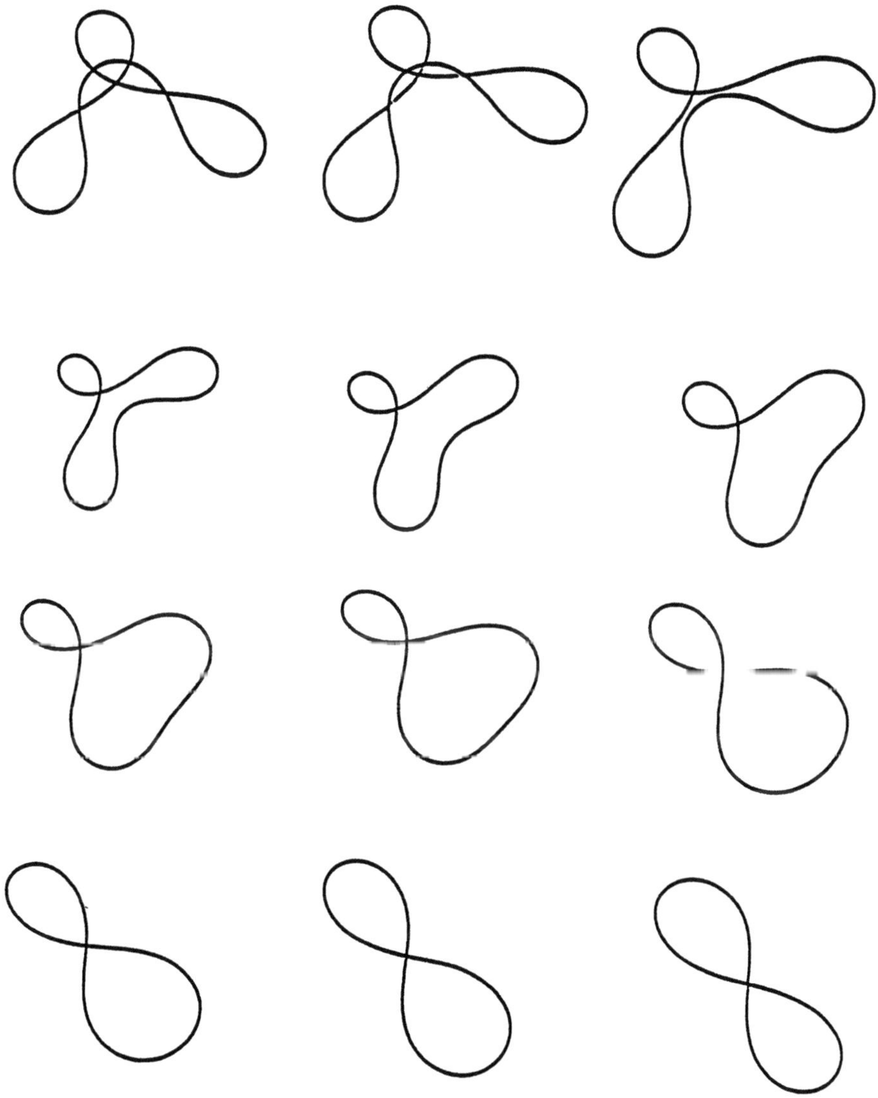

Figure 3.2

REFERENCES

1. Horn, B. K. P., *The curve of least energy*, ACM Trans. Math. Softw. **9** (1983), 441–460.
2. Langer, J. and Singer, D. A., *Curve straightening and a minimax argument for closed elastic curves*, Topology **24** (1985), 75–88.
3. Langer, J. and Singer, D. A., *Curve-straightening in Riemannian manifolds*, Ann. Global Anal. Geom. **5** (1987), 133–150.
4. Linnér, A., *Some properties of the curve straightening flow in the plane*, Trans. Amer. Math. Soc., Vol. 315, Nr. 1, September 1989 (1988).
5. Rudin, W., "Real and Complex Analysis," Third Edition, McGraw–Hill, pp. 145–148, 1987.
6. Abraham, R.; Marsden, J. and Ratiu, T., "Manifolds, Tensor Analysis and Applications," Addison–Wesley, 162 pages, 1983.

Department of Mathematics and Statistics, Case Western Reserve University, Cleveland, OH 44106

Geometric Data for Triply Periodic Minimal Surfaces in Spaces of Constant Curvature

KONRAD POLTHIER

Abstract. In this note we describe the use of geometric data for the construction of triply periodic minimal surfaces in R^3, S^3 and H^3. With a conjugate surface construction we obtain the Plateau solution of a fundamental piece for the symmetry group of the minimal surface. For some examples in R^3 a method of H. Karcher and M. Wohlgemuth has led to the Weierstraß formula.

Introduction.

Triply periodic minimal surfaces (TPMS) in a three space $\bar{M}^3(c)$ of constant curvature c are defined to be periodic in three independent directions. As a common characteristic of the TPMS described in the following we have that a fundamental piece (Color Plate 9) for the symmetry group sits in a polyhedron meeting the faces at a right angle along the minimal surface boundary. To get a quick impression of the shape of the complete minimal surface it is sometimes more convenient to combine a few fundamental pieces to a fundamental cell. This cell also sits in a polyhedron which should usually be quite regular.

The first example was discovered by H. A. Schwarz about 1865 and independently by B. Riemann. The fundamental cell sits in a regular cube with holes to all faces of the cube and it has all the cubical symmetries. By successive translation along the edges of the cube one obtains a complete, embedded TPMS in R^3 (Color Plate 10).

During the last years these TPMS have been of interest in cristallography and chemistry because they may be seen in relation with the spatial pattern of atoms and bondings in certain crystal structures and to phase transitions. Most interesting among the TPMS are those which are complete and embedded, since they divide $\bar{M}^3(c)$ into two disjoint regions, which form two interpenetrating labyrinths separated by the minimal surface. One labyrinth of the Schwarz surface is inside the cells connected by the holes on the faces, the other labyrinth is outside. Because of the straight lines on this minimal surface one can interchange both labyrinths by rotating one by 180° around the line. So both labyrinths are congruent and the cell bisects the cube's volume (Color Plates 10,12).

A large collection of new TPMS in R^3 was published by A. Schoen in a 1970 NASA Report [**7**] (Color Plates 9,13,14). He described 12 additional minimal surfaces and created from these and others with the help of experimental data beautiful plastic models. His names describe characteristics of the space lattices of the two labyrinths. The existence of these surfaces and many new examples was proved by H. Karcher [**2**] using a conjugate surface construction. In cases in which planar symmetry lines cut the surfaces in quadrilateral pieces excellent pictures were obtained by D. Andersson. These surfaces are graphs in tetrahedral coordinates and he solved the equations $H = 0$ and $H =$ constant numerically.

H. Karcher and M. Wohlgemuth [**9**] found a method to obtain the Weierstraß data from TPMS in R^3 if the symmetry lines are suitable, the Gauß map is well-known and the genus of the underlying Riemann surface is low (Color Plates 9,13,14).

In 1986 TPMS were obtained by H. Karcher, U. Pinkall and I. Sterling [**3**] in S^3 and recently by H. Karcher and K. Polthier [**6**] in H^3 using a modified conjugate surfaces construction. The fundamental cells of these surfaces sit in regular polyhedra which arise from ordinary space tesselations (Color Plate 11).

Symmetry Properties.

Minimal surfaces have a very exquisite symmetry property: if they contain a straight line, then rotation around this line leaves the surface invariant, and if they contain a planar geodesic one can reflect the surface in that plane. These properties follow from the Schwarz reflection principle of function theory. It was generalized by Lawson [**4**] to minimal surfaces in spaces of constant curvature.

Let

$$F : M^2 \to \bar{M}^3(c)$$

be a minimal immersion of a simply connected piece of a Riemann surface M^2 into a space $\bar{M}^3(c)$ of constant curvature c. Then, according to Lawson, we get the associate family of F:

Let ds be the metric and S be the Weingarten map of F then there exists an isometric family

$$F_\theta : M^2 \to \bar{M}^3(c), \quad 0 \leq \theta \leq 2\pi$$

of minimal immersions with geometric data

$$ds = ds_\theta$$

$$S_\theta = D^\theta * S$$

where D^θ is the rotation by θ in the oriented tangent space of M^2. F_θ is the associate family of F and $F_{\pi/2}$ is called the conjugate immersion to F.

In this notation the Frenet data for a curve $F_\theta c$ for the frame $(\partial F_\theta \dot{c}, N, \partial F_\theta(D^{\pi/2}\dot{c}))$ on the minimal surface is given by

$$\text{(1)} \qquad \begin{aligned} \text{curvature}(c) &= k_\theta = \langle S_\theta \dot{c}, \dot{c} \rangle = -\tau_{\theta+\pi/2} \\ \text{torsion}(c) &= \tau_\theta = \langle S_\theta \dot{c}, D^{\frac{\pi}{2}} \dot{c} \rangle = k_{\theta+\pi/2} \end{aligned}$$

for a geodesic c. This means for geodesics

$$\begin{aligned} F \cdot c \text{ is a straight line } &\Leftrightarrow k = \langle S \cdot \dot{c}, \dot{c} \rangle = 0 \\ &\Leftrightarrow \tau_{\pi/2} = 0 \\ &\Leftrightarrow F_{\pi/2} \cdot c \text{ is a planar curve.} \end{aligned}$$

Therefore a straight line is a planar curve in the conjugate immersion and vice versa a planar geodesic becomes a straight line.

Existence Via a Conjugate Surface Construction.

In cases in which planar symmetry lines cut the minimal surface into simply connected pieces, the boundary of the conjugate piece consists of linear lines which are in R^3 orthogonal to the symmetry planes. Using (1), the rotation of the normal along a symmetry arc is the same as the rotation of the tangent plane along the conjugate straight segment. Since both pieces are isometric the angles in corresponding vertices are the same. In R^3 in many cases of A. Schoen's and H. Karcher's minimal surfaces, this determines the conjugate boundary configuration; sometimes one has to apply an intermediate value argument to a one-parameter family of configurations. If the conjugate boundary has a monoton projection onto the boundary of a convex planar domain we have the existence of a unique Plateau solution bounded by the straight lines, and conjugation leads to the desired patch.

Using this method, B. Smyth [8] showed the existence of three disk-type, embedded minimal surfaces in every tetrahedron in R^3 meeting the boundary of the tetrahedron at a right angle. This shows existence in cases in which the fundamental piece of A. Schoen's minimal surfaces is bounded by four planar curves.

The fundamental cells of the periodic minimal surfaces of H. Karcher, U. Pinkall and I. Sterling **[3]** in S^3 and H. Karcher and K. Polthier **[6]** in H^3 sit in regular (p,q,r)-polyhedra with holes to all plane faces similar to the Schwarz surface in a cube (Color Plate 11). Dividing a polyhedron by its planes of symmetry in a fundamental tetrahedron with dihedral angles $\frac{\pi}{2}, \frac{\pi}{2}, \frac{\pi}{2}, \frac{\pi}{p}, \frac{\pi}{q}, \frac{\pi}{r}$ would divide the minimal surface cell in a fundamental patch with four planar boundary curves. The existence of this patch can also be shown with a conjugate surface construction, but the argumentation becomes more involved than in R^3, since one cannot directly infer the desired tetrahedron for the space tesselation from the quadrilateral. The angles of the quadrilateral are four of the six dihedral angles of the desired tetrahedron, namely $\frac{\pi}{2}, \frac{\pi}{2}, \frac{\pi}{2}$ and $\frac{\pi}{q}$. The two free sidelength parameters must be chosen in such a way that the other two dihedral angles are $\frac{\pi}{p}$ and $\frac{\pi}{r}$. Control over the construction comes from the maximum principle, helicoids are used as comparison surfaces.

Deformations of TPMS in R^3 to triply periodic constant mean curvature surfaces with the same symmetry group have been constructed by H. Karcher **[2]**. Using Lawson **[4]**, to the Plateau solution of a spherical polygon in S^3 corresponds a conjugate path bounded by S^3-planar geodesics which lead to a constant $= 1$ mean curvature patch in R^3 bounded by R^3-planar geodesics. It remains to determine the Plateau contours in S^3 such that the corresponding H-piece in R^3 generates the desired symmetry group in R^3.

Weierstraß Formula for TPMS in R^3.

Let $F : M^2 \to R^3$ be a minimal immersion of a compact Riemann surface into R^3. Then there exists the Weierstraß representation of F

$$F(z) = \operatorname{Re}\left\{\int^{Z}\left(\frac{1}{g} - g,\, i * \left(\frac{1}{g} + g\right),\, 2\right) * g\eta\right\} \in R^3$$

with a meromorphic function g and a meromorphic differential η. Multiplying the integrand with $e^{i\varphi}$, $\varphi \in [0, 2\pi)$ leads to the associate family.

Since η is a little bit unwieldy, H. Karcher and M. Wohlgemuth introduced a globally defined meromorphic function μ with

$$g\eta =: \mu \frac{dg}{g}.$$

This special choice of μ will reduce computations and simplify symmetry considerations. The geometric data expressed in terms of g and μ are:

$$ds = \left(\frac{1}{|g|} + |g|\right) * |\mu| * \left|\frac{dg}{g}\right|$$

$$\langle Sw, w\rangle_z = 2 \operatorname{Re}\left\{\mu(z) * \left(\frac{dg(w)}{g(z)}\right)^2\right\} \text{ with } z \in M,\ w \in T_zM.$$

Parametrizing with the Gauß map, $\frac{1}{|g|} + |g|$ and $\left|\frac{dg}{g}\right|$ are invariant under reflections across a meridian and the equator lines. Therefore, if $|\mu|$ is also invariant under reflection across a meridian or the equator, this line must be a geodesic.

Looking for the symmetry lines it will also suffice to examine μ. Let c be a geodesic, then

$$\begin{aligned} c \text{ is straight } &\Leftrightarrow k(c) = \langle S\dot{c}, \dot{c}\rangle = 0. \\ &\Leftrightarrow \mu(c) * \left(\frac{dg(\dot{c})}{g(c)}\right)^2 \in i * R. \\ c \text{ is planar } &\Leftrightarrow \tau(c) = \langle D^{\pi/2} S\dot{c}, \dot{c}\rangle = 0. \\ &\Leftrightarrow \mu(c) * \left(\frac{dg(\dot{c})}{g(c)}\right)^2 \in R. \end{aligned}$$

Since for meridians and equator arcs we always have $\left(\frac{dg}{g}\right)^2 \in R$, $\mu \in R(iR)$ indicates a planar (straight) symmetry line on a geodesic meridian or equator arc.

Riemann surface theory states that for two meromorphic functions g and μ on a compact Riemann surface exists a polynomial equation

$$P(g, \mu) = 0$$

with $\deg_g P = \deg \mu$ and $\deg_\mu P = \deg g$. In the other direction, if we now guess such a polynomial equation for g and μ such that g and μ satisfy the desired symmetry conditions of the minimal surface, then this equation indeed defines a Riemann surface.

The remaining problem is to guess the polynomial equation. Similarly as in the case of minimal surfaces with ends one finds a first relation between g and μ from the metric formula: since we want no ends of the surface which would arise from a pole of $|ds|$ and no branchpoints of the metric ($= |ds| = 0$), we have

$$0 < |ds| < \infty$$

as a necessary condition. Therefore g and μ mutually have to compensate zeroes and poles. This dependence can be best shown in a tableau:

g	μ
0^k	0^{k+1}
∞^k	0^{k+1}
a^k	∞^{k-1}

where 0 stands for zeroes, ∞ for poles and a for branchpoints of g. No further zeroes and poles of μ are allowed.

A similar tableau explaining the dependence of g and η was used by H. Karcher and M. Callahan, D. Hoffman, W. Meeks **[1]** in the case of minimal surfaces with ends. The degree of g (deg g) can be read from the minimal surface and deg μ can then be obtained from the tableau.

This strategy and a "a lot of good guessing" (W. Meeks) has led to the Weierstraß formulae for a number of minimal surfaces of A. Schoen and H. Karcher. The R^3-TPMS in this article were computed using those Weierstraß formulae.

REFERENCES

1. M. Callahan, D. Hoffman and W. Meeks, *Embedded minimal surfaces with an infinite number of ends*, Preprint, Amherst (1987).
2. H. Karcher, *The triply periodic minimal surfaces of Alan Schoen and their constant mean curvature compagnions*, Preprint SFB256 (1988).
3. H. Karcher, U. Pinkall and I. Sterling, *New minimal surfaces in* S^3, Preprint, Max-Planck-Institut für Mathematik, Bonn (1986).
4. H. B. Lawson, Jr., *Complete minimal surfaces in* S^3, Ann. of Math. **92** (1970), 335–374.
5. K. Polthier, Calendar, *Bilder aus der Differentialgeometrie, 1989*, Vieweg-Verlag (1988).
6. K. Polthier, *Neue Minimalflächen in* H^3, Diplomarbeit, Bonn (1989).
7. A. Schoen, *Infinite periodic minimal surfaces without selfintersections*, Technical Note D-5541, NASA, Cambridge, Mass. (1970).
8. B. Smyth, *Stationary minimal surfaces with boundary on a simplex*, Invent. Math. **76** (1984), 411–420.
9. M. Wohlgemuth, *Abelsche Minimalflächen*, Diplomarbeit, Bonn (1988).

Department of Mathematics, University of Bonn, Beringstr. 4, 5300 Bonn, West Germany

Embedded Triply-Periodic Minimal Surfaces and Related Soap Film Experiments

ALAN H. SCHOEN

Abstract. Some aspects of the modelling of embedded triply-periodic minimal surfaces (ETPMS) by both soap films and plastic models are discussed. Eight new examples of possible ETPMS are introduced. Interference-diffraction patterns obtained by reflecting laser light from soap films with non-zero Gaussian curvature are described.

1. Introduction.

When I learned in 1966 about the several examples of embedded triply-periodic minimal surfaces (ETPMS) which were investigated by Schwarz (Schwarz [**1**]) in 1866 and later by his student Neovius (Neovius [**2**]), I wondered what other surfaces of this type might either be known or waiting to be discovered. A year later, while at the short-lived NASA Electronics Research Center in Cambridge, I began to learn a little about the mathematics of minimal surfaces. I found that by combining elementary information about minimal surfaces, space groups, and polyhedral packings with some heuristic ideas about dual infinite graphs and their associated 'labyrinths', I could derive some plausible ETPMS candidates. In a few cases, I was able to prove embeddedness. A summary of what I had found by the end of 1969 is contained in a NASA Technical Note (Schoen [**3**]).

Probably every mathematician knows that no small part of the pleasure of studying minimal surfaces derives from their esthetic appeal. Schwarz and his colleagues made extraordinary plaster models of ETPMS (Neovius [**2**]). I felt early on that I also needed some physical models to look at. I made rather crude models of the ETPMS of Schwarz in 1966, by taping together thin vinyl surface modules made with a toy vacuum-forming machine. The surface moulds for vacuum-forming were made by stretching rubber sheet material over a metal frame and then pouring either liquid plaster or polyester resin into the assembly. Two years later I met the sculptor Harald Robinson, who taught me how to use 'shrink wrap' film in place of rubber sheet, and epoxy for the casting material. Harald made most of the moulds used to fabricate the models illustrated in Schoen [**3**].

After NASA/ERC was dissolved in 1970, I became less active in the study of minimal surfaces. I did not stop looking for examples of ETPMS, however. Eight of the new cases I found in 1970–1975 are described in §5. Until 1975, I continued to think about the problem from time to time, and sometimes I used an experimental method, employing a laser, which I had developed at NASA for investigating hypothetical ETPMS (Harald Robinson performed all the early experiments of this type for me). Some of these surfaces can be loosely described as obtained from known examples of ETPMS by cutting holes out of the surfaces and attaching cylindrical tubules at the hole rims, and others by cutting tubules out of the surface and filling in the holes left behind. This laser method is described in §3.

I was intrigued by reports, in 1968–1970, of a variety of both natural and man-made objects which appear to mimic the shapes of particular ETPMS. These objects include:

(i) the interface between the elemental constituents of certain 'segregated' binary metal alloys, like Au-Ni, which appear at low temperatures ('spinodal' transformation first treated by Willard Gibbs);
(ii) the microtubular ultrastructure (or prolamellar body) of the cells of the leaf membranes of green plants which have been exposed to prescribed sequences of light and darkness;
(iii) the arrangement of the lipid molecules of certain anhydrous soaps, grown as single crystals, at elevated temperatures; and
(iv) the surface of the single crystal calcite skeleton of certain microscopic marine animals.

2. Construction of plastic models of ETPMS based on soap film experiments.

In order to make a mould for the vacuum-forming of plastic replicas of a minimal surface module, it is first necessary to determine the geometry of the boundary of the module. If the boundary edges are all straight line segments, this is a straightforward problem, because the orientation of the edges is known. Their relative lengths may be to some extent prescribed—because of embeddedness—by crystallographic constraints. On the other hand, if one or more of the boundary edges is curved (line of curvature in a plane of reflection symmetry), and the Weierstrass parametrization of the surface is unknown, one can resort to soap films to model the boundary.

A relatively simple experimental method for determining the shape of a curved boundary edge is to photograph the corresponding edge of the appropriate soap film in a stationary state inside a transparent polyhedral box. The walls of the box coincide with planes of reflection symmetry of the extended surface. Corners of the box are removed where necessary, to allow the insertion of a soda straw, so that one can blow on the soap film and thereby move it into a position which corresponds to a stationary state. If there is only one curved edge per face of the enclosing box, then so long as a nylon string is stretched along any straight line which lies in the interior of the surface, the soap film is found to be in stable equilibrium. Without such a string, the film is not in stable equilibrium, but it will remain near its stationary state position for at least a second or two because of viscous drag along the interior walls of the enclosing box.

A mould frame is now assembled from thin metal plates—one edge of each plate defining an edge of the surface module—to model the boundary of the surface module. Next a sheet of thin plastic 'shrink wrap' (e.g., Cryovac from W. R. Grace Co.) is draped over the mould frame, taped to the edge plates, and then judiciously shrunk into an approximation of the required minimal surface by short blasts from an industrial heat gun (or large hair dryer). When the surface shape of the plastic film is judged to be sufficiently accurate, a thin coating of epoxy is applied to the outside of the film in order to stiffen it. After this coating has hardened, the interior surface of the film is sprayed with an anti-stick compound, and then the interior volume of the mould frame (the space between the base of the mould frame and the interior surface of the film) is filled with epoxy. Finally, the metal plates of the mould frame are removed, and the mould is complete. Replicas of surface modules can now be fabricated by standard vacuum-forming techniques.

After the vacuum-formed surface modules are cut out of the plastic sheets from which they are formed, they are joined along their edges with long-lasting transparent adhesive tape. Surface coloring can be used to good advantage, especially if one uses white polystyrene plastic sheet for the surface modules, applying silk-screened color to one side of the plastic surface (before the vacuum-forming is carried out).

When the boundary of the surface module consists of curved edges and is derived by the soap-film-in-a-box method described above, capillarity effects in the neighborhood of the boundary of the soap film limit the accuracy of the derived boundary shape. I investigated the error associated with this effect in each case studied, by measuring the length of each of the curved boundary edges and—making use of the fact that the image of each boundary curve in the *adjoint* minimal surface is a straight line segment—computing the amount by which the chain of these computed straight edges fails to close.

In those cases where the soap film (inside the polyhedral box) is a model of a minimal surface which contains a 2-fold axis of rotational symmetry, I modelled the surface both with and without a nylon string stretched inside the box along the 2-fold axis. There was always an observable difference between these two results. With the nylon string in place, the amount by which the computed adjoint boundary failed to close was satisfactorily small, especially when the number of boundary arcs of the minimal surface did not exceed the number of faces of the polyhedron by more than one. In the absence of the nylon string, the discrepancy was larger, but even with the string in place, if the number of boundary arcs exceeded the number of polyhedral faces by two or more, the amount by which the computed adjoint boundary failed to close was sometimes appreciable, indicating that the conjectured embedded surface does not exist. That was disappointing, because these examples include hypothetical ETPMS whose Riemann surfaces have interestingly large genus (as high as 39). I believe that some of these high genus surfaces may be authentic minimal surfaces, but to obtain persuasive experimental evidence for their existence, it would be necessary to resort to the method described in §3. This method provides an experimental test for the existence of embedded ETPMS whose elementary surface modules are bounded by mirror-symmetric plane lines of curvature. It also makes possible an improvement in the accuracy of the determination of the shapes of the curved edges of a surface module, and therefore it was used for the design of the mould frames for all curved-edge surface modules which do not contain straight lines.

When using the soap-film-in-a-box method, it is important to choose the size of the box to be small enough to minimize the distorting effects of gravity on the inscribed soap film, and at the same time large enough to make the free energy associated with the area of the film significant compared to the free energy associated with the boundary edges, which are far from ideally thin.

Color Plates 5–8 show examples of ETPMS, in the form of plastic models.

3. Laser-goniometer method for deriving the shapes of the curved edges of a surface module.

This method exploits two properties of adjoint minimal surfaces:

(a) the direction of the normal vector at corresponding points of two adjoint surfaces is invariant (same Gauss map); and

(b) the straight boundary edges of one of the adjoint surfaces are orthogonal to their respective curved edge images, lying in planes of reflection symmetry, of the other.

Let S be the curved-edge surface module which is to be modelled, and S' its straight-edged adjoint. S' can be modelled by a soap film spanned by a boundary composed of straight edges (tightly stretched monofilament nylon makes good edges). The boundary frame is supported by a vertical lead screw near the center of a goniometer table, with the edge along which the surface orientation measurements are to be made aligned vertically. The direction of the normal to the soap film surface is measured at a number of points near this edge. After each such measurement, the surface is displaced by the lead screw through a measured distance parallel to the edge, and then the surface is rotated about the edge so that a horizontally incident laser beam is retro-reflected from the surface precisely along the optic axis.

The measurements are slow and tedious but straightforward. Soap solutions for producing the necessary long lasting films are readily available. A simple computer program converts the angle vs. distance measurements into computer plots of the curved edges. These plots can then be used for shaping the edge plates of a vacuum-forming mould frame used to fabricate surface modules.

4. Derivation of surfaces with one or more undetermined parameters (edge length ratios).

For a curved-edge minimal surface module S, so long as there is only one curved edge incident on each face of the enclosing polyhedron K (kaleidoscopic cell), the relative lengths of the straight edges of the adjoint surface module S' are constrained only by the requirement that they define a closed polygon. If S has two edges e_i and e_j incident on one face of K, then it is necessary somehow to determine the unique value for the ratio of the lengths of e_i and e_j which will make the extended surface an embedding.

This is known as the vanishing period problem in analytic treatments of such surfaces. Experimentally, this problem can be solved by a method of successive approximations, as I shall now explain by means of a specific example.

In Schoen [**3**] (pp. 72–73), it is mentioned that it is in some ways useful to classify some of the examples of ETPMS that have reflection symmetries according to a description based on the shape of a multiply-connected surface contained in a suitable space-filling polyhedral cell. For example, the Schwarz surface P is composed of multiply-connected surface modules which can be described loosely as obtained by puncturing six holes in a central sphere inside an enclosing cube, and then attaching a cylindrical tubule to the rim of each puncture hole and also to the center of an appropriate cube face. In another ETPMS (called I-WP), eight tubules extend from the central sphere to the corners of an enclosing cube.

Now let us consider a 'hybrid' surface, called O,C-TD (Schoen [**3**], pp. 28–30). There is one tubule incident on each of the six cube faces and one on each of the eight cube corners. The straight-edged boundary of a module of the *adjoint* of O,C-TD—let us call it A(O,C-TD)—is interpolated between modules of the respective adjoints of P and I-WP as follows (cf. Figure 1):

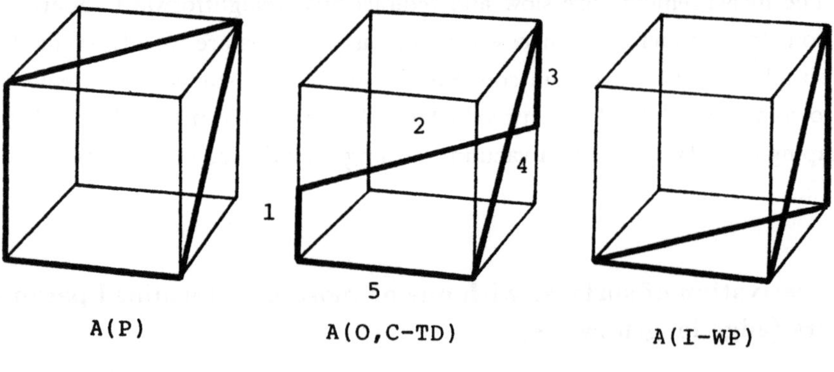

Figure 1

The boundary of a module of A(O,C-TD) is defined by a circuit of five directed edges. Two non-consecutive edges (1 and 3), which are separated by a single edge (2), are parallel. If we steadily lengthen edge 1 in small increments, while simultaneously shortening edge 3 at the same rate, leaving the lengths of edges 2, 4, and 5 unchanged, we eventually obtain a circuit which is congruent to the boundary of A(P) (this occurs precisely when edge 3 vanishes). If on the other hand we steadily lengthen edge 3 while shortening edge 1 at the same rate, we eventually obtain a circuit which is congruent to the boundary of A(I-WP) (this occurs precisely when edge 1 vanishes).

The boundary of A(O,C-TD) is obtained by interpolation between the boundaries of A(P) and A(I-WP). The intermediate value theorem implies that there exists an interpolated boundary for a surface whose adjoint—O,C-TD—is embedded. The correct value for the relative lengths of edges 1 and 3 in A(O,C-TD) insures that in O,C-TD, the (curved) adjoint images of these two edges will be coplanar, and not merely in parallel planes (Schoen [**3**], pp. 45–46).

The case of O,C-TD required the determination of the value of one parameter—the ratio of the lengths of edges 1 and 3 in A(O,C-TD). An experimental value for this ratio was derived by successive approximations in a series of laser-goniometer measurements on four boundary frames of different proportions. In 1969–1974, four other such 1-parameter hybrid surfaces and a 2-parameter surface which is not of hybrid type were also derived by this method. In the latter case, measurements were made on six differently proportioned adjoint boundary frames before a satisfactory solution was obtained. This ETPMS can be described loosely as constructed from a module, enclosed in a cube, which consists of a central spherical chamber with tubules extending to the centers of the upper front and rear cube edges, and also to the centers of the lower left and right cube edges.

In another set of experiments, what seemed to be a plausible hypothetical ETPMS was postulated, but the laser measurements of surface normals indicated that no such embedded surface exists. This surface can be described loosely as a network of tubules which enclose all the edges of an infinite packing of truncated octahedra.

5. Eight examples of possible ETPMS.

In 1970, I discovered that a pair of identical double rings, each in the shape of a figure-eight formed by two tangent circles of the same diameter, can be used to span an annular soap film, by using the same technique one uses to make a soap film model of a catenoid spanned by a pair of single rings. This suggested the possibility of making a modification of the Schwarz surface P, by substituting squares for the circles of the figure-eights. (The P surface itself can be generated by applying Schwarz's reflection principle at the straight edges of the annular surface spanned by two squares of edge length 1 which are separated vertically by a distance equal to $\sqrt{2}/2$.) The shape of the required wire frames is shown in Figure 2a. Even though the boundaries are self-intersecting polygons, all horizontal sections of the surface which lie between the two boundaries are simple closed curves. Repeated application of Schwarz's reflection principle to the edges of this annular surface and its images yields an ETPMS.

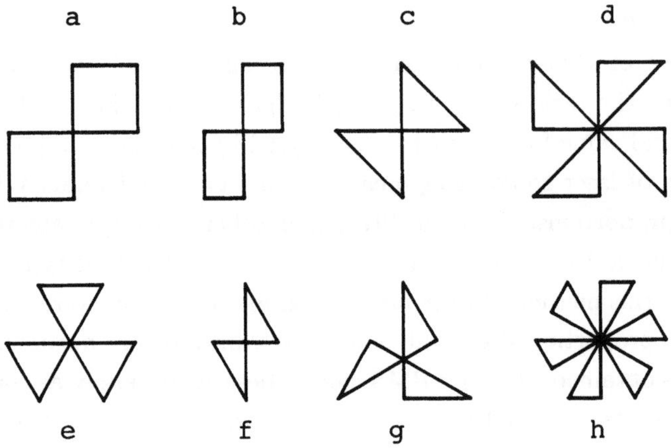

Figure 2

One quickly discovers that in order to make a soap film model of this annular surface, it is necessary to perform an extra operation which is not required with single square rings: when the frames are separated after being withdrawn from the soap solution, there is an extraneous vertical soap film which is parallel to the long axis of either frame. It is necessary to rotate this film through a quarter-turn about the vertical direction, by blowing on it with a straw, before removing it (by puncturing).

When I asked Fred Almgren in 1974 how one would prove the existence of the minimal surface modelled by this annular soap film, he suggested using barrier theory. I didn't know anything about barrier theory, and I didn't pursue the matter. A few months later, it dawned on me that if this surface could be proven to exist, then there must also be seven other annular surfaces of this type which lead to ETPMS. I soldered pairs of wire frames in the shapes shown in Figure 2 for these additional surfaces, and I found that all of these surfaces require the two-step rotation-puncture removal of extraneous vertical films. My students and I constructed epoxy moulds and papier-mache models of the surfaces corresponding to Figures 2a and 2c. The genus of the Riemann surfaces for the ETPMS derived from these boundary frames is found from the Gauss–Bonnet formula to be 5,5,9,9,13,13,13,13, for Figures 2a–2h, respectively.

6. Non-localized interference-diffraction patterns produced by reflecting a laser beam from soap films with non-zero Gaussian curvature.

Early in 1974, I once again assembled a laser, goniometer, and lead-screw (cf. §3) for investigating the existence of additional examples of 'hybrid' ETPMS (cf. §4). I immediately discovered that the graininess in the red spot reflected onto the white screen displayed considerable structure. To examine this structure, I needed only to enlarge the spot by moving the target screen farther from the soap film.

Figure 3a shows a typical example of the patterns observed with this experimental arrangement. A 1 mw He-Ne laser beam was reflected from the region of greatest curvature of a soap film spanned by a symmetrical saddle boundary wire. The principal radius of curvature in this part of the soap film was approximately 1 cm. Figure 3b shows an example of the circular fringes produced by a spherical soap bubble of diameter 3 cm. (The intensity modulations in the smaller ring patterns of Figure 3b resemble the Airy fringes produced by diffraction by a small circular aperture.) These photographs were taken with a camera with its lens removed, located about 30 cm from the soap film. Because the interference patterns are non-localized, they are 'in focus' everywhere, and no lens is necessary. No interference patterns of this type are observed with films which are plane or in the shape of a right circular cylinder. It was found that only when the Gaussian curvature of the film is non-zero is the reflected light sufficiently widely dispersed over the area of the target screen to produce an observable pattern.

By examining a number of plane soap films under a microscope, I concluded that in every freshly prepared soap film, small droplets—presumably of lenticular shape—nucleate by precipitating out of the solution (and then

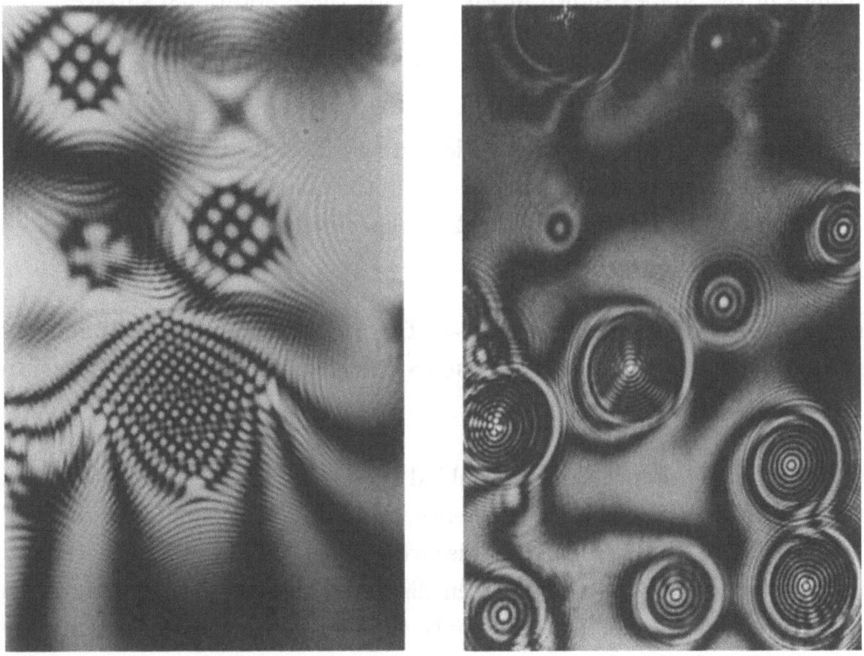

a b

Figure 3

glide around the the film under the influence of the Marangoni effect until shortly before the film 'wrinkles' and breaks). I identified the reticular grid pattern units, like those visible in Figure 3a, with reflection from single droplets. These pattern units are observed to undergo continuous transformations. This suggested to me that when the order of the grid (defined as the square root of the number of its interstices) increases or decreases by one, the droplet has glided to a place where the film thickness has changed by a half-wavelength of the incident light.

Detailed measurements of the spacing of the nearly perfect hyperbolic fringes in the simplest observed patterns ('degenerate' grids, which I assumed were produced by reflection from the smallest lenticular droplets as they approached critical size), suggested an analogy with Newton's rings, the circular fringes produced by reflection from the front and rear surfaces of the space between a spherical and a flat glass surface in contact. The central portion of one such hyperbolic fringe pattern is visible in the upper portion of Figure 3a. Quite isolated examples of such patterns, in which a large number of undistorted fringes are visible, are frequently observed. The nearly perfect hyperbolic shape of these fringes suggested to me that a hyperbolic paraboloid would be a sufficiently accurate model of a minimal surface in the neighborhood of a droplet, but I failed to come to a definite conclusion about the extent to which internal reflections from back surfaces contribute to the reflected pattern, and I did not develop a detailed mathematical model of the effect.

In January 1975, I showed my results to Joseph Keller, who quickly developed an analysis of the hyperbolic interference fringes produced by a plane wave which is reflected both from the front spherical surface of a droplet and also from the front surface of the hyperbolic paraboloid film in which it is embedded. Because of my procrastination, none of this work—including Keller's analysis—has ever been published. I hope that Keller will soon publish his analysis, and that others may be encouraged by this tardy report to investigate the problem further.

References

1. H. A. Schwarz, "Gesammelte Mathematische Abhandlungen," Vol. 1, Julius Springer, Berlin, 1890.
2. E. R. Neovius, "Bestimmung Zweier Speciellen Periodische Minimalflächen," Helsingfors, 1883.
3. A. H. Schoen, *Infinite periodic minimal surfaces without self-intersections*, NASA Technical Note TN D-5541 (1970).

Department of Electrical Engineering, Southern Illinois University, Carbondale, IL 62901

The Collapse of a Dumbbell Moving Under Its Mean Curvature

J. A. SETHIAN

Abstract. The collapse of a dumbbell moving under its mean curvature has attracted considerable attention. A new class of numerical algorithms has been developed recently that can follow hypersurfaces propagating with curvature-dependent speed in any number of space dimensions. The essential idea behind these algorithms is to view the propagating hypersurface as a particular level set of a higher dimensional function. The motion of this higher-dimensional function is described by a scalar Hamilton–Jacobi equation with parabolic right-hand-side. This equation may be easily solved using techniques borrowed from the solution to hyperbolic conservation laws. We demonstrate this technique applied to the problem of a dumbbell collapsing under its own mean curvature. Our results show the breaking of the handle, the developing singularity, and the collapse of the two separate sections.

I. Introduction: Motivation and Statement of Problem.

The need to follow fronts moving with curvature-dependent speed arises in the modeling of a wide class of physical phenomena, such as crystal growth, flame propagation and secondary oil recovery. In each of these problems, the motion and structure of the interface depends on an intricate feedback mechanism between local effects due to the front itself, such as curvature and tangential stretch, and global effects in the underlying media, such as heat diffusion and boundary conditions. The complexities of the propagating boundaries can include sharp cusps where the curvature becomes singular, and topological changes, such as breaking, merging and extinction. All told, this poses a formidable challenge to numerical algorithms.

Recently (see [**6**]), we have developed a class of numerical algorithms that can follow hypersurfaces propagating with curvature-dependent speed in any number of space dimensions. The essential idea behind these algorithms is to view the propagating hypersurface as a particular level set of a higher-dimensional function. In this higher-dimensional space, sharp cusps and topological changes are handled quite naturally with no special

Supported in part by the Applied Mathematics Subprogram of the Office of Energy Research under contract DE-ACO3-76SF00098, NSF under the National Science Foundation Mathematical Sciences Program, and the Sloan Foundation.

consideration. The motion of this higher dimensional function is described by a scalar Hamilton–Jacobi equation with parabolic right-hand-side. This equation may be easily solved using techniques borrowed from the solution to hyperbolic conservation laws.

We now state the problem. We wish to follow the evolution of an initial surface propagating along its gradient field with speed $F(K)$ a given a function of the curvature (either mean or Gaussian in more than one space dimension). The outline of this paper is as follows. First, we discuss traditional numerical algorithms to the problem. Then we present the Hamilton–Jacobi level set formulation. Finally, we give some examples, and study in detail the collapse of a dumbbell.

II. Traditional Numerical Techniques.

Most numerical techniques to approximate the equations of motion for this problem fall into one of two categories:

A. Marker Particles.

The first set of techniques rely on a parameterization of the moving front. This parameterization is then discretized into a finite set of marker points. The normal direction to the curve is determined by discrete derivatives of the marker positions. Curvature and stretch are also approximated by discrete derivatives of the marker positions. To move the front, the positions of the marker points are updated in time according to approximations to the equations of motion. Such techniques can be quite accurate in the attempt to follow the motions of small perturbations for short times. For details of an involved technology using this approach, see [**8**].

However, for large, complex motion, several problems soon occur. First, marker particles come together in regions where the curvature of the propagating front builds. When this happens, the computed curvature can change drastically from one particle to the next because of small errors in the positions. Consequently, oscillations soon develop in the front which grow unavoidably. Since the size of the time step to insure stability is controlled by the spacing between marker points, a prohibitively small time step is required. The usual solution is a regridding mechanism to redistribute marker particles so that they remain a minimum distance apart, allowing a practical time step. Unfortunately, this regridding mechanism usually contains an error term which resembles diffusion and dominates the real effects of curvature under analysis, see [**9,10**]. Thus, one has chosen to sacrifice the most interesting propagation characteristics, such as front sharpening and curvature singularities, in order to keep the calculation alive. Time and effort are often spent solving an unrelated problem.

Topological changes and higher dimensionality are even bigger obstacles for marker particle schemes. Suppose two separate patches merge and their boundary becomes a single curve. It is difficult to produce a systematic way of removing those markers that no longer sit on the actual boundary. The bookkeeping of removing, redistributing, and reconnecting marker particles becomes even more complicated for higher dimensional interface problems.

B. "Volume of Fluid" Techniques.

The second set of techniques, called "volume of fluid" techniques, track the motion of the interior region rather than the boundary, see [**1,2,5**]. In these algorithms, the interior is discretized, usually by employing a grid on the domain and assigning to each cell a "volume fraction" corresponding to the amount of interior fluid currently located in that cell. An advantage of such techniques is that no new computational elements are required as the calculation progresses (unlike the parametrization methods), and complicated topological boundaries are easily handled. The front is moved by constructing local polygonal approximations to the front in each cell, based on the neighboring volume fractions. Unfortunately, it is difficult to calculate front properties, such as curvature and normals from such a crude representation of the boundary.

To summarize, (1) marker particle methods suffer from instability and topological limitations because they follow a local representation of the front, and (2) volume of fluid techniques sacrifice most of the interesting information about a front in exchange for a crude approximation that handles topological changes and stability requirements.

III. Hamilton–Jacobi Algorithms.

In this section, we present a different formulation of the equations of motion that is particularly amenable to numerical approximation. This formulation was first presented in [**6**].

A. Formulation of Equations of Motion.

Let $\gamma(t)$ be the position and front at time t. We view the evolving front $\gamma(t)$ as the level set of a higher-dimensional function Ψ. To be more precise, let the initial surface $\gamma(0)$ be a closed, non-intersecting hypersurface of dimension $N-1$. We construct the function Ψ by letting $\phi(\bar{x},0) = \pm d$, $\bar{x} \in R^N$, where d is the distance from $\bar{x}$ to $\gamma(0)$, with the plus (minus) sign chosen if $\bar{x}$ is inside (outside) $\gamma(0)$. We now require a time-dependent differential equation for Ψ corresponding to the evolution of $\gamma(t)$. Since each level set propagates in its normal direction with speed $F(K)$, the equation for the evolving surface is

$$\Psi_t - F(K)\,|\nabla\Psi| = 0,$$

which is reminiscent of a Hamilton–Jacobi equation. Note that

1) Ψ is a function of $R^N \times [0,\infty) \to R$, thus we have added an extra dimension to the problem.
2) At any time t, the position of the front $\gamma(t)$ is just the level set of all $\bar{x}$ such that $\Psi(\bar{x},t) = 0$.

This is an Eulerian formulation of the front propagation problem. The level surface $\Psi = 0$ may change topology as it moves, either breaking into multiple parts or fusing together. For any fixed t, slicing Ψ by the level plane at height 0 retrieves the position of the front. As illustration, in Figure 1 we show an expanding circle in R^2 and the corresponding surface evolving in R^3.

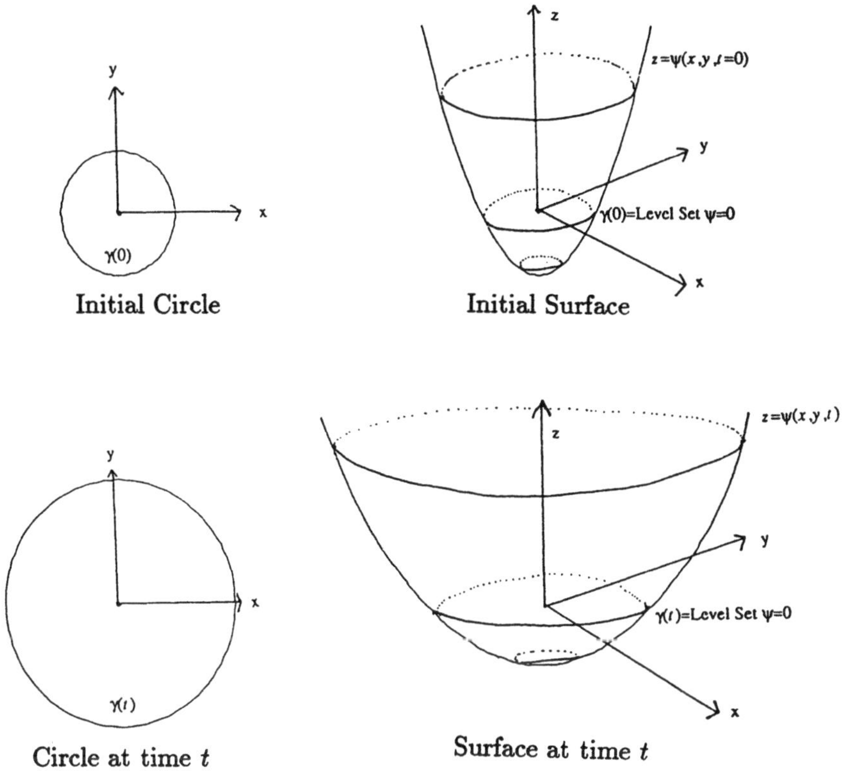

Figure 1

B. Numerical Algorithms Based on Hyperbolic Conservation Laws.

Together with the initial position of the front, the above equation is an initial value partial differential equation. We solve this initial problem by directly exploiting the technology from schemes to approximate hyperbolic conservation laws. As a motivation to understand this connection, consider the initial value Hamilton–Jacobi equation

$$\phi_t - (1 + \phi_x^2)^{1/2} = 0$$

where $x \in R$ and $\phi : R \times [0, \infty) \to R$. This is a simplified version of our equation of motion, and applies in the case when the front moves at constant speed and remains a function $\phi(x,t)$ for all time. If we differentiate with respect to t and let $u = \Psi_x$, we have an equation for the propagating slope, namely

$$u_t + [G(u)]_x = 0$$

where $G(u) = -(1+u^2)^{1/2}$. This is a hyperbolic conservation law which may be solved by a variety of methods, all carefully designed to handle discontinuities in the solution. These discontinuities in the slope u correspond to sharp corners in our propagating fronts. The success of these techniques lies in an adequate numerical flux function g which approximates the flux $G(u)$. Given this numerical flux function g, we work directly with the Hamilton–Jacobi equation and write

$$\Psi_j^{n+1} = \Psi_j^n + \Delta t \cdot g.$$

These schemes can be generalized to handle more complicated speed functions $F(K)$ and several space dimensions. For details, see [**6**,**7**].

IV. Examples.

As illustration, we use these techniques to approximate the solution to three problems. In Figure 2, we show a wound spiral moving inwards with speed equal to its curvature. Gage [**3**] showed that a convex curve must shrink to a point as it flows under its curvature. Recently, Grayson [**4**] has shown that any non-intersecting curve must collapse smoothly to a circle under this motion. In Figure 2a-d we show the collapse to a circle and eventual disappearance at $t = .295$. (The surface vanishes when $\Psi_{ij}^n < 0$ for all ij.) For details, see [**6**].

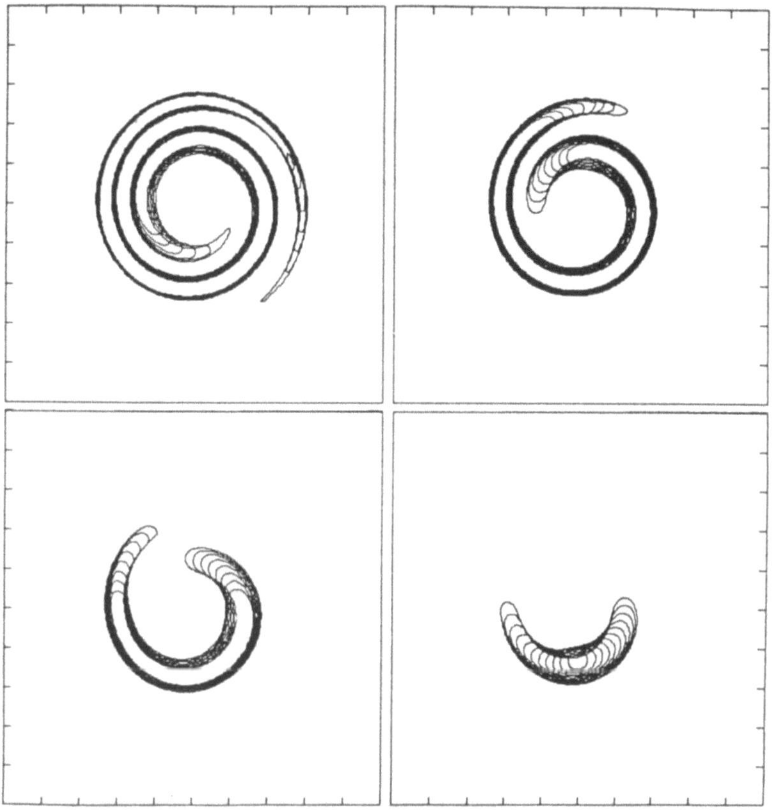

Figure 2: Spiral Collapsing Under Its Own Curvature: $F(K) = -K$

In Figure 3, we show the burning of a flame in the shape of a torus, separating products on the inside from reactants outside, and burning with speed $F(K) = 1 - .1K$, where K is the mean curvature. First, the torus burns smoothly (and reversibly) until the main radius collapses to zero. At that time ($T = 0.3$), the genus goes from 1 to 0, characteristics collide, and entropy condition is automatically invoked. The surface then looks like a sphere with deep inward spikes at the top and bottom. These spikes open up as the surface moves, and the surface approaches the asymptotic spheroidal shape. For details, see [**6**].

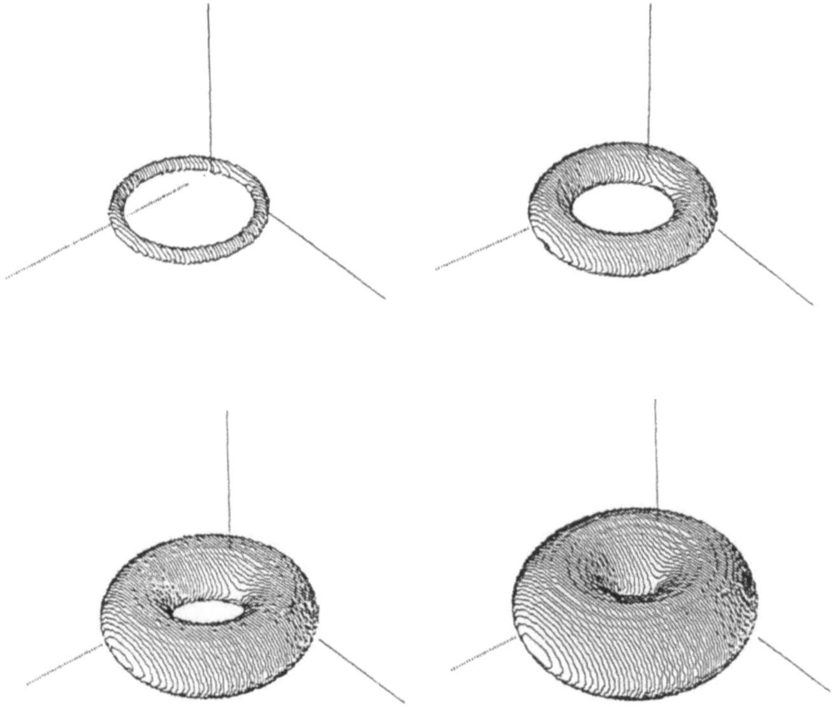

Figure 3: Expanding Torus
$F(K) = 1 - \epsilon K, \quad \epsilon = .01, \quad T = 0.0, 0.1, 0.2, 0.3$

Finally, we study the collapse of a dumbbell. Consider the dumbbell made up of two spheres, each of radius .3, and connected by a cylindrical handle of radius .15. The x axis is the axis of symmetry. We wish to follow the evolution of this surface as it moves under its mean curvature.

We must be slightly careful in setting up the problem. Our algorithm computes the curvature at each mesh point by evaluating the expression for the mean curvature. Both the numerator and denominator vanish at the center of this dumbbell. Evaluation of this ratio causes the computer code to halt. Formally speaking, whenever this occurs we should insert the correct limiting form of the expression $F(K)\nabla\Psi$ into our algorithm. Unfortunately, the correct limiting form is a point of controversy, and one wants to avoid building into the code an *a priori* assumption about the behavior of the figure where the singularity occurs. We circumvent this problem by using an even number of grid points in all three coordinate directions x, y, and z, so that points are staggered around the center axis of rotation. We performed a calculation with 214 grid points in the x direction, and 72 grid points in both the y and z directions. The computational box stretched from -1.0 to 1.0 in the x direction, and $-1/3$ to $1/3$ in both the y and z directions. We chose a time step of $\Delta t = .0002$.

The results of our calculation are shown in Figure 4. We show a diagonal cross-section of the dumbbell (that is, the intersection of the moving surface with the plane $y = z$). Although the initial shape is only piecewise continuous, the corners are immediately smoothed out as the surface moves inward. The position of the front is plotted every 100 time steps until the handle becomes small, and then every 10 time steps. The figure shows the narrowing of the handle as the surface shrinks, and the break into two distinct pieces each of which collapses to a point. For details, see [**11**].

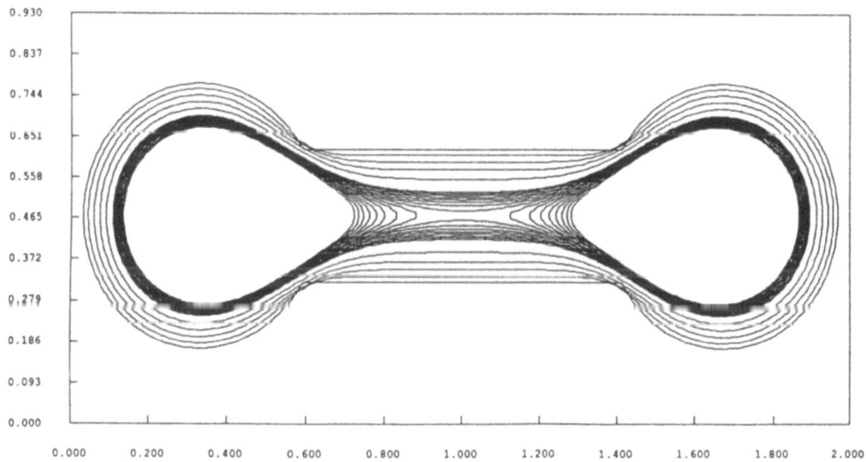

Figure 4: $F(K) = -K \quad 214 \times 72 \times 72$ Grid

An interesting issue is the type of singularity (i.e., a corner or a cusp) that develops. It is not clear from our preliminary calculations which occurs as the topology changes. This problem is under study, and we expect to report on the results elsewhere.

References

1. Chorin, A. J., *Curvature and solidification*, J. Comput. Phys. **58** (1985), 472.
2. ———, *Flame advection and propagation algorithms*, J. Comp. Phys. **35** (1980), 1.
3. Gage, M., *An isoperimetric inequality with applications to curve shortening*, Duke Math. J. **50** (1983), 1225.
4. Grayson, M., *The heat equation shrinks embedded plane curves to round points*, J. Diff. Geom. **26** (1987), 285.
5. Noh, W. and Woodward, P., in Proceedings, Fifth International Conference on Fluid Dynamics, A. I. van de Vooran and P. J. Zandberger, Eds., Springer–Verlag (1976).
6. Osher, S. and Sethian, J. A., *Fronts propagating with curvature-dependent speed: algorithms based on Hamilton–Jacobi formulations*, J. Comp. Phys. **79** (1988), 12.
7. Osher, S. and Sweby, P. K., "Recent Developments in the Numerical Solution of Nonlinear Conservations Laws," State of the Art in Numerical Analysis, A. Iserles and K. W. Morton, Eds., Cambridge Press, 1986.
8. Overman, E. A. and Zabusky, N. J., *Contour dynamics—an interface method for studying the evolution of large density gradient ionospheric plasma clouds*, in "Fronts, Interfaces and Patterns," Proceedings of the Third International Conference of the Center for Nonlinear Studies, North Holland, Amsterdam (1984).
9. Sethian, J. A., *Curvature and the evolution of fronts*, Comm. Math. Phys. **101** (1985), 487.
10. ———, *Numerical methods for propagating fronts*, in "Variational Methods for Free Surface Interfaces," edited by P. Concus and R. Finn, Springer–Verlag, New York (1987).
11. ———, *Recent numerical algorithms for hypersurfaces propagating with curvature-dependent speed: Hamilton–Jacobi equations and conservation laws*, to appear, J. Diff. Geometry, November 1989.

Department of Mathematics and Lawrence Berkeley Laboratory, University of California, Berkeley, CA 94720

Geometry Versus Imaging: Extended Abstract

ALVY RAY SMITH

Geometry Versus Imaging.

There are two quite distinct ways of making pictures with computers. The geometric way is quite widely understood—and often thought to be the only way. The imaging way is less intuitive—and leads to a different marketplace which is probably as large or larger than that for geometry. The terminologies, theories, and even heroes of the two worlds are quite distinct, and the hardware devices to implement them are strikingly different.

Figure 1 illustrates the two domains and their interrelationships. Geometry-based picturing begins with the description of objects or scenes in terms of common geometric ideas—polygons, lines, spheres, cylinders, patches, splines, etc. Recall that these are mathematical abstractions, not pictures. To make a digital picture of a geometrically described object requires that it be *rendered* (or *rasterized* or *scan converted*) into pixels. Geometric concepts live in real continuous space, requiring floating point for accurate computer representation. Famous names are Pythagoras and Euclid. The theorems of analytic geometry are of paramount importance.

Imaging-based picturing begins with a set of discrete samples—pixels—of a continuum, usually placed on a uniform grid. As Figure 1 shows, these samples *may* come from scan conversion of geometry, but in general they do not. In the majority of cases they come from non-geometric sources such as digitized satellite photographs, computed tomographic (CT) or magnetic resonance (MR) medical scans, digitized X-radiographs, electronic paint programs, seismic sensors, supercomputer simulations of partial differential equation systems, or laboratory measurements. In all cases, an array of numbers (samples) is the original data—not a "display list" of geometric primitives. Imaging generates pictures from this data by directly displaying them on the computer screen. The imaging domain is discrete by definition and integer arithmetic typically suffices. Famous names are Nyquist and Fourier. The Sampling Theorem is of paramount importance.

†The full text of this paper can be found in *Computer Graphics World*, November, 1988.

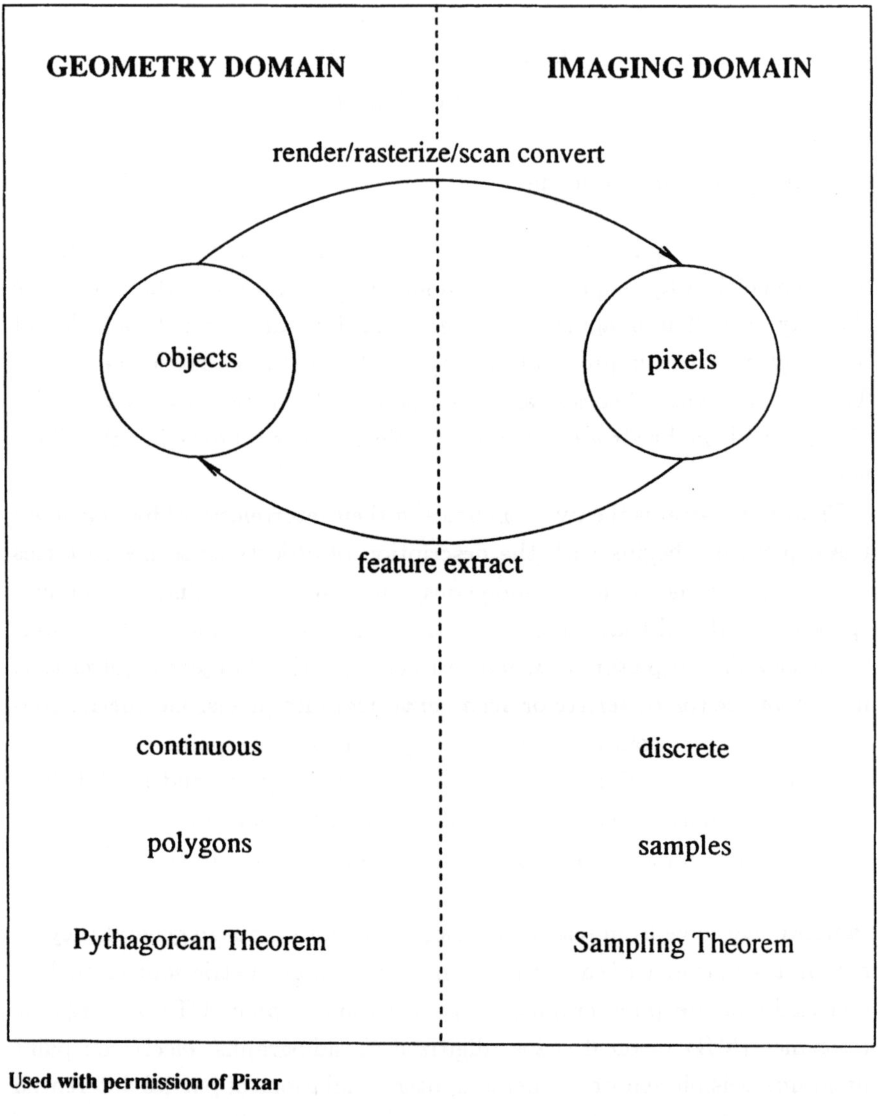

Used with permission of Pixar

Figure 1

As Figure 1 also points out, it is possible in some cases to extract geometric data from sampled data and reenter the geometric domain (and then render the geometry to reenter the image domain!). This step is not required, however, to make pictures—contrary to the opinion of a surprisingly large number of people. In fact, it frequently introduces thresholding artifacts (jaggies) which may be highly undesirable—as in medical diagnostic imaging which insists, and depends, on no alterations of its data. Direct imaging of data arrays avoids such artifacts.

Notice that the distinction between geometry and imaging is *not* that between image synthesis and image analysis. An electronic paint program is an excellent example of a non-geometric synthesis technique—an imaging technique. Nor is it the distinction between "computer graphics" and "image processing". Image processing is only a subset of imaging, and computer graphics loosely covers picturing from both domains. The fundamental difference is whether the elemental datum is a geometric or a numeric entity—a polygon or a pixel.

Architectural Differences.

The geometry and imaging distinction is reflected in special-purpose hardware *accelerators* available for each. All graphics computations discussed here *could* be implemented on a general-purpose computer, such as a workstation host, minicomputer, mainframe, or personal computer. But in the late 1980s, it is still the case that these offer insufficient price/performance for geometry and imaging computations. The general-purpose machines with relatively lower computational power simply cannot do the computations in a tolerable amount of time or lack sufficient memory; the more powerful machines use cycles which are too expensive compared to what can be purchased for much less in accelerators.

Geometry accelerators are measured in terms of the number of geometrical objects they can manipulate in realtime. Imaging datasets are typically so large that realtime is not yet an appropriate measure for them (unless programmability is sacrificed). So imaging accelerators are measured in terms of the number of pixels they can comfortably manipulate on the order of 100 times faster than a host computer. Some geometry engines do a little imaging, and some imaging computers do a little geometry. The purpose of the next section is to clarify the distinctions.

GEOMETRY SOPHISTICATION METER

Shading "Shade"	Geometry "Shape"	Anti-aliasing	Complexity #Primitives Per Picture
shading language			
materials			
shaped lights			
distributed lights			
matte/glossy			
displacement maps			
environment maps			
bump maps			
image texture maps			
radiosity			
refraction			
procedural textures			
Phong shading	hyperpatches	motion	100,000,000
transparency	patches	textures	10,000,000
Gouraud shading	quadrics	specular	1,000,000
multiple lights	nurbs	edges	100,000
flat shading	polygons	lines	10,000
none	lines	none	1,000

Used with permission of Pixar

Geometry sophistication increases up each column, from bottom to top.

The state of realtime geometry machines in 1988 is below the lines in each column. The RenderMan Interface addresses the entire chart with particular emphasis above the lines.

Figure 2

IMAGING SOPHISTICATION METER

Techniques	Dimensions	Filtering	Complexity #Pixels Per Image
3D painting, FFT, and compositing			
volumetrics			
volume imaging			
warping			
classification: thematic, MR, and CT			
FFT, Walsh, and other transforms			
compression			
soft-edged painting and soft fill		very wide	
convolutions and		bessel	1,000,000,000
filters		sinc	100,000,000
histograms and	volume movies	cubic	10,000,000
equalization	volumes	gauss	1,000,000
point operations	image movies	box	100,000
matte algebra	images	none	10,000

Used with permission of Pixar

Imaging sophistication increases up each column, from bottom to top.

This chart measures imaging accelerators, but some *geometry* accelerators do limited imaging - below the lines in each column. Image computers can address the entire chart.

Figure 3

Sophistication Meters.

Figures 2 and 3 are *sophistication meters* for geometry and imaging respectively. They are attempts to summarize the major techniques and terms in the two domains. Both charts are ordered from bottom to top in each column by increasing sophistication. The horizontal lines in the columns mark the *realtime lines* for each chart. These are placed somewhat generously and are explained more fully below. Notice that the terms used on the two charts are almost completely different.

Geometry Sophistication Meter. The capabilities of realtime geometry accelerators in 1988 lie below the realtime lines of Figure 2. The most obvious observation is that most of what is known about geometry-based graphics has not yet been pulled into realtime. This is particularly true of the "shading" of geometric objects—their visual content—which is in general more difficult than the shaping of the objects—their geometry. The recently proposed RenderMan Interface is exactly a roadmap to all the non-realtime geometry-based graphics.

Imaging Sophistication Meter. The first thing to notice about Figure 3 is that sophisticated imaging applications are tremendously complex in terms of pixel count. This is the world addressed by imaging computers. Some geometry accelerators confusingly offer restricted imaging capabilities as well as geometry. The realtime lines in this chart indicate where in the scheme of things this limited imaging lies. In particular, it is restricted to essentially 1.25K $\times$ 1K display memories. Of course, the host computer in geometry workstations can always do the imaging—by definition of computing—but then it executes at general-purpose price/performance, not the order-of-magnitude lower price/performance of a special-purpose imaging accelerator. This is standard "you get what you pay for" general-purpose computing.

Pixar, 3240 Kerner Boulevard, San Rafael, CA 94901

Constant Mean Curvature Tori

I. STERLING

Constant mean curvature tori in $\mathbf{R}^3$ were first discovered, in 1984, by Wente [**15**]. These examples solved the long standing problem of Hopf [**6**]: Is a compact constant mean curvature surface in $\mathbf{R}^3$ necessarily a round sphere? Hopf proved that if the surface is topologically a sphere then it must be round and Alexandrov [**3**] proved that if the surface is embedded then it must be a round sphere.

We will abbreviate the term "constant term curvature $H = \frac{1}{2}$" by writing "CMC."

In 1985, Abresch [**1**] classified all CMC tori having one family of planar curvature lines. These surfaces are given explicitly in terms of elliptic integrals (by solving a system of ODEs) and include "Wente tori". Walter [**13**] gives even more explicit representations of these tori in terms of elliptic and theta functions of Jacobi type. Spruck [**12**] shows that the original method of Wente yields exactly those tori in Abresch's classification.

New examples, "twisted tori," were found by Wente [**16**] and studied by Abresch [**2**]. Abresch showed that these new examples also arise by solving a certain system of ODEs.

Recently Hitchin [**5**] has reduced the problem of finding all harmonic maps $T^2 \to S^3$ to a problem in algebraic geometry. His construction gives as a special case all possible Gauss maps $T^2 \to S^2 \subset S^3$ of CMC tori.

Finally we remark that Kapouleas [**7**] has found compact constant mean curvature surfaces for every genus ≥ 3.

In the author's joint work with U. Pinkall [**10**] we prove that Abresch's ODE-system corresponds to the case $n = 1$ of an infinite family of similar systems defined for each $n \in \mathbb{N}$. In general these systems describe quasiperiodic immersions $F : \mathbf{R}^2 \to \mathbf{R}^3$ with constant mean curvature. We call these surfaces "of (finite) type n". Surfaces of finite type are fairly dense among all CMC surfaces: At a non-umbilic point every CMC surface can be approximated to any prescribed order of differentiation by a surface of finite type.

We prove [**10**] that all immersions $F : \mathbf{R}^2 \to \mathbf{R}^3$ covering CMC tori are necessarily of finite type and we provide an efficient criterion to single out CMC tori among general surfaces of finite type. In this sense we have constructed *all* CMC tori in $\mathbf{R}^3$. Our algorithm for finding CMC tori can be implemented on a computer. In [**10**, §8] we discuss in detail the cases $n = 1, 2$ and present computer pictures of a two-parameter family of CMC tori with $n = 2$.

We now give an outline of our method. Let $\Omega \subset \mathbf{R}^2 = \mathbf{C}$ be a domain, $F : \Omega \to \mathbf{R}^3$ a conformal parameterization by curvature lines of a CMC surface (all this will henceforth be abbreviated by saying that F is a CMC immersion). Then, granted suitable normalizations, it is well-known that the metric $g = 4e^{2\omega}ds^2$ induced on Ω via F satisfies

$$\omega_{z\bar{z}} + \frac{1}{2}\sinh(2\omega) = 0. \tag{1}$$

In Eq. (1) we have used the Cauchy-Riemann operators

$$\frac{\partial}{\partial z} = \frac{1}{2}\left(\frac{\partial}{\partial x} - i\frac{\partial}{\partial y}\right), \quad \frac{\partial}{\partial \bar{z}} = \frac{1}{2}\left(\frac{\partial}{\partial x} + i\frac{\partial}{\partial y}\right),$$

so $\omega_{z\bar{z}} = 1/4\,\Delta\omega$, Δ the Laplacian of $\mathbf{R}^2$.

Conversely, given any metric g as above such that (1) holds, there is an isometric CMC immersion of (Ω, g) into $\mathbf{R}^3$. In this sense the theory of CMC surfaces in $\mathbf{R}^3$ without umbilic points is equivalent to the study of the "elliptic sinh-Gordon equation" Eq. (1).

The key ingredient in our approach comes from studying infinitesimal deformations of CMC surfaces, i.e., the tangent space to the (infinite dimensional) manifold of CMC immersions $F : \Omega \to \mathbf{R}^3$. This has two aspects, an intrinsic and an extrinsic one. From the intrinsic viewpoint one is led to consider the "linearized sinh-Gordon equation"

$$\dot{\omega}_{z\bar{z}} + \cosh(2\omega)\dot{\omega} = 0. \tag{2}$$

From the extrinsic viewpoint infinitesimal deformations of CMC surfaces (by CMC surfaces) are given by normal vector fields $\dot{F} = uN$ (N the unit normal, $u : \Omega \to \mathbf{R}$ a function) such that deforming F in the direction of $\dot{F}$ preserves the mean curvature H to first order. In this case we call $\dot{F}$ a Jacobi field along F. It turns out that $\dot{F} = uN$ is a Jacobi field along a CMC immersion F if and only if u solves the "Jacobi equation"

$$u_{z\bar{z}} + \cosh(2\omega)u = 0. \tag{3}$$

It is convenient in Eq. (3) to allow u to be complex-valued. Since Eq. (3) is a real equation, u solves Eq. (3) if and only if both Re u and Im u solve Eq. (3).

Now, while deforming F in the direction $\dot{F} = uN$, the corresponding ω also changes, and in [**10**, §2] we develop an algorithm that allows us to compute the infinitesimal change $\dot{\omega}$ of ω from the function u. This, together with the simple observation that Eqs. (2) and (3) are the *same* PDE, leads to the following iterative procedure: Start with any solution of Eq. (3), for example with $u_1 = \omega_z$. Use u_1 to perform an infinitesimal normal deformation of the given surface and obtain as the corresponding change of ω a new solution $\dot{\omega}$ of Eq. (2). Then again $\dot{F} = u_2 N$ with $u_2 = \dot{\omega}$ will be a Jacobi field etc. This procedure yields for any given solution u_1 of Eq. (3) an infinite sequence $u_1, u_2, \ldots$ of such solutions. In the special case where $u_1 = \omega_z$ we are able to calculate the whole sequence $u_1, u_2, \ldots$ explicitly as polynomials in the partial derivatives of ω.

Suppose now that ω is doubly periodic with respect to some lattice Γ in $\mathbf{R}^2$ (ω being doubly periodic means that the corresponding surface in $\mathbf{R}^3$ is pieced together by congruent fundamental domains; this surface covers a torus only if certain closing conditions are satisfied). Then ω as well as $u_1, u_2, \ldots$ can be considered as functions on the torus $T^2 = \mathbf{R}^2/\Gamma$. Since the elliptic Eq. (3) has only a finite dimensional solution space there must be a linear combination of the form

$$u_{n+1} = \sum_{\ell=1}^{n} a_\ell u_\ell + b_\ell \bar{u}_\ell. \tag{4}$$

Since all u_ℓ are differential polynomials of ω, Eq. (4) is a second PDE (in addition to (1)) which ω satisfies. More generally, ω is called a "solution of finite type" if Eq. (4) holds with suitable $a_\ell, b_\ell \in \mathbf{C}$.

Equation (4) insures that all partial derivatives of ω of order $2n+1$ can be expressed in terms of derivatives of lower order. Pursuing this further we finally obtain two commuting vector fields Z and $\bar{Z}$ (corresponding to the z- and $\bar{z}$-derivatives in $\mathbf{R}^2$) on the $(2n)$-order jet space of ω. In this way we accomplish the reduction of the elliptic sinh-Gordon equation (1) to a system of ODE's in the doubly periodic case (and more generally for solutions of finite type).

In Ref. [**10**, §6] we show that every CMC immersion $F : \mathbf{R}^2 \to \mathbf{R}^3$ of finite type has an "axis", i.e., there is a fixed straight line in $\mathbf{R}^3$ that can be computed just from the local data of the surface. The direction of this axis is given by a certain constant vector field $J : \mathbf{R}^2 \to \mathbf{R}^3$ whose coefficients with respect to the local moving frame of our surface can be computed from the partial derivatives of ω.

To exploit all the information contained in the vector field J along F we use the fact that every solution ω of Eq. (1) gives rise to a whole one-parameter family F_θ of CMC surfaces in $\mathbf{R}^3$, which depend on an angular parameter θ (all surfaces in this family are intrinsically isometric). This family is called the "associated family" of CMC surfaces for the CMC immersion $F = F_o$. The coefficients of the corresponding vector fields J_θ with respect to the local moving frame turn out to be Fourier polynomials in θ. Expanding $\langle J_\theta, J_\theta \rangle$ as a Fourier polynomial P in θ we obtain as Fourier coefficients many first integrals of our ODE-system (recall that $J_\theta(z)$ is independent of z). Moreover, using standard methods from solution theory [**4**], we use the Fourier polynomial P to define a certain hyperelliptic Riemann surface M. We embed our ODE-system into the Jacobian of M in such a way that the flows become linear. This gives us full control over all the periodicity properties of our ODE-system and finally leads to a criterion to single out CMC tori among general CMC surfaces of finite type.

The elliptic sinh-Gordon equation $\Delta\omega + 2\sinh(2\omega) = 0$ belongs to a whole family of similar equations which are obtained if we replace sinh by sine or cosh and perhaps the Laplacian Δ by the wave operator $\partial^2/\partial x^2 - \partial^2/\partial t^2$ in the x-t-plane. From the complex viewpoint all these equations are equivalent, but the different real forms can have quite different properties.

For the hyperbolic sine-Gordon equation the special solutions corresponding to our solutions of finite type are usually called "finite gap solutions" or "n-soliton wavetrains." They are obtained from an ODE-system in C_*^{6n-1} very similar to our ODE-system. A crucial problem (referred to as the "reality problem") remains to parametrize explicitly the set of those initial conditions in C_*^{6n-1} that give rise to real solutions $\omega : \mathbf{R}^2 \to \mathbf{R}$. Ercolani and Forest have solved this problem completely for the case $n = 1$ [**4**]. We solved the reality problem for the elliptic sinh-Gordon equation by starting with the ODE-system mentioned above (no reality problem there) and mapping them into C_*^{6n-1}. We expect that our methods should apply also to the hyperbolic sine-Gordon equation.

With minor modifications all results in [**10**] apply also to CMC-surfaces with arbitrary mean curvature H in the 3-sphere S^3 and to CMC-surfaces with $|H| > 1$ in hyperbolic space. All such surfaces are related to solutions of the sinh-Gordon equation. By the maximum principle there cannot be compact CMC-surfaces in H^3 with $|H| \leq 1$ (or in $\mathbf{R}^3$ with $H = 0$). Nevertheless our methods can be used to study complete surfaces of this type. They depend on solutions of the cosh-Gordon equation ($|H| < 1$ in H^3) or the Liouville equation ($H = 1$ in H^3 or $H = 0$ in $\mathbf{R}^3$).

To find CMC-tori in S^3 is easier than in $\mathbf{R}^3$: Given an (intrinsically) doubly periodic surface in S^3 its holomony group lies in $SO(2) \times SO(2)$, so only four rationality conditions on the rotation angles are needed to make the surface close up. This contrasts to the case of $\mathbf{R}^3$ where the translational periods had to be zero. Hitchin [**5**] has constructed all harmonic maps $T^2 \to S^3$ (and among these all minimal immersions) from suitable algebro-geometric data. Our results for the case of S^3 differ from Hitchin's in the sense that the data needed for Hitchin's construction involve reality conditions, whereas our method needs only rationality conditions in order to yield closed tori.

So to a large extent all CMC tori in S^3 are now known. This knowledge should be helpful in studying the following conjecture, which generalizes the well-known Hsiang–Lawson conjecture that the only embedded minimal torus in S^3 is the Clifford torus.

CONJECTURE: Every embedded CMC-torus in S^3 is a torus of revolution.

References

1. Abresch, U., *Constant-mean curvature tori in terms of elliptic functions*, J. reine u. angew. Math. **374** (1987), 169–192.
2. ———, *Old and new periodic solutions of the sinh-Gordon equation*, Seminar on new results in nonlinear partial differential equations, Vieweg (1987).
3. Alexandrov, A. D., *Uniqueness theorems for surfaces in the large*, I Vestnik Leningrad Univ. Math. **11** (1956), 5–17.
4. Ercolani, N. M. and Forest, M. G., *The geometry of real sine-Gordon wavetrains*, Comm. Math. Phys. **99** (1985), 1–49.
5. Hitchin, N. S., *Harmonic maps from a 2-torus to the 3-sphere*, Preprint (1987).
6. Hopf, H., *Differential geometry in the large*, Lecture Notes in Mathematics **1000**, Springer (1983).
7. Kapouleas, N., *Constant mean curvature surfaces in Euclidean three space*, Bull. AMS, Vol. 17, No. 2 (October 1987).
8. Moser, J., *On the theory of quasiperiodic motions*, SIAM Review, Vol. 8, No. 2 (April 1966), 145–172.
9. Mumford, D., "Tata Lectures on Theta II," Birkhäuser, 1984.
10. Pinkall, U. and Sterling, I., *On the classification of constant mean curvature tori*, to appear in Annals of Math.
11. Simons, J., *Minimal varieties in Riemannian manifolds*, Ann. of Math. **88** (1968), 62–105.
12. Spruck, J., *The elliptic sinh-Gordon equation and the construction of toroidal soap bubbles*, Preprint.
13. Walter, R., *Explicit examples to the H-problem of Heinz Hopf*, Geom. Dedicata **23** (1987), 187–213.
14. Warner, F., "Foundations of Differentiable Manifolds and Lie Groups," Scott and Foresmann, 1971.
15. Wente, H. C., *Counterexample to a conjecture of H. Hopf*, Pacific Jour. of Math., Vol. 121, No. 1 (1986), 193–243.
16. ———, *Twisted tori of constant mean curvature in* $\mathbb{R}^3$, Seminar on new results in non-linear partial differential equations, Vieweg (1987).

Department of Mathematics, Tech. Univ. of Berlin, (01000) Berlin (12), Fed. Rep. Germany

On Deformable Models

DEMETRI TERZOPOULOS

Abstract. Deformable models are a new class of physically-based modeling primitives for use in computer graphics, especially for animation. Deformable curves, surfaces, and solids are governed by the mechanical principles of continuous bodies whose shapes can change over time. These primitives are capable of a variety of behaviors, including elasticity, viscoelasticity, plasticity, fracture, conductive heat transfer, thermoplasticity, melting, and fluid-like behavior in the molten state. By numerically simulating the equations of motion that govern deformable models, while visualizing their state variables, we are able to create realistic animations depicting nonrigid objects in simulated physical worlds.

1. Introduction.

Methods to describe shape and motion are of central concern to computer graphics modeling. Computational geometry has proven useful for modeling stationary objects whose shapes do not change over time [**1**]. For computer animation, however, we have found it tremendously beneficial to employ concepts from computational physics. Physically-based animation offers unparalleled realism compared to conventional animation techniques [**2**]. It consists in applying simulated forces to models that inhabit simulated physical worlds including constraints, obstacles, heat sources, and a wide variety of environmental influences. Numerical procedures compute the model's reactions, as dictated by Newtonian mechanics, while images are synthesized by graphically rendering the state variables though time.

This paper reviews a physically-based technique that we have developed for modeling nonrigid shape and motion. *Deformable models* are a class of curve, surface, and solid modeling primitives founded on principles governing the dynamics of nonrigid bodies [**3**]. The models are governed by differential equations of motion that include deformation force terms. The terms are expressed as variational derivatives of deformation energy functionals constructed using differential-geometric measures of shape. Although computationally more demanding than most free-form geometric primitives, deformable models are superior in several ways. Most importantly, since deformable models are fundamentally dynamic, they provide a unified description of shapes and their motions through space.

Our physically-based modeling work has a distinctly different emphasis than engineering analysis. Deformable models are designed to facilitate the creation of visually realistic computer animations. To this end, they idealize broad regimes of material response under rather diverse environmental conditions. Deformable models are capable of a wide variety of natural behaviors, including elasticity, viscoelasticity, plasticity, fracture, conductive heat transfer, thermoplasticity, melting, and fluid-like behavior in the molten state. Their parameters describe qualitatively familiar behaviors, such as stretchability, flexibility, resiliency, fragility, conductivity, viscosity, etc. Deformable models are appropriate in computer graphics applications, where a keen concern with tractability and convenience motivates mathematical abstraction and computational expediency.

Color Plates 27–30 present a selection of synthetic images created by numerically integrating the equations of motion associated with deformable models in response to the forces at work in simulated physical worlds. Color Plates 27a–c show a deformable membrane model shrink wrapping around a rigid jack which acts as an impenetrable geometric constraint. Color Plate 28 shows a frame taken from an animation of a deformable flag model waving in a simulated wind. Color Plate 29 shows a physically-based manipulation of a deformable face model made of "computational plasticine." Color Plate 30 shows a deformable sheet model that was sheared by simulated forces.

2. Elastic Models.

Elastically deformable models recover their reference shapes completely as soon as all applied forces causing deformation are removed. We express the internal shape restoring forces in terms of deformation potential energies. A generalization of the ideal spring, the elastically deformable model stores potential energy during deformation, which it then releases as it recovers the reference shape.

We describe two types of elastically deformable models: a *primal formulation* [4], which explicitly represents the configuration of the deformable model in space, and a *hybrid formulation* [5], which decomposes this configuration into a deformation component and a reference component that undergoes rigid-body motion.

The primal formulation handles free motions implicitly using a non-quadratic deformation energy functional (nonlinear restoring forces). This formulation is suitable for moderately to highly flexible behavior, but the numerical conditioning of the discrete equations deteriorates with increasing rigidity. The hybrid formulation permits the use of a quadratic deformation energy functional (linear restoring forces). Although less appropriate for highly deformable models, the equations of motion offer a significant practical advantage for fairly rigid models, especially when complex reference shapes are desired. Numerical conditioning improves as the model becomes more rigid, tending in the limit to well-conditioned, rigid-body dynamics. Thus, the primal and hybrid formulations complement one another in practice, and together they cover a wide range of deformable behaviors from highly elastic to nearly rigid.

In both formulations $\mathrm{u} = (u_1, \ldots, u_d)$ will denote the material coordinates of points in a body domain $\Omega = [0,1]^d$, where d is the dimensionality of the domain: $d = 1$ for curves, $d = 2$ for surfaces and $d = 3$ for solids.

2.1. Primal Formulation.

The primal formulation describes deformations using the positions

$$\mathbf{x}(\mathrm{u}, t) = [x_1(\mathrm{u}, t), x_2(\mathrm{u}, t), x_3(\mathrm{u}, t)]^T \tag{1}$$

of mass elements in a body relative to an inertial frame of reference Φ in Euclidean 3-space, where the subscripts 1, 2, and 3 denote the X, Y, and Z axes, respectively, of the reference frame.

A deformable model is described by the positions $\mathbf{x}(\mathrm{u}, t)$, velocities $\partial\mathbf{x}/\partial t$, and accelerations $\partial^2\mathbf{x}/\partial t^2$ of its mass elements as functions of the material coordinates u and time t. Lagrange's equations of motion for $\mathbf{x}$ in the inertial frame Φ take the form [**6**]:

$$\mu\frac{\partial^2\mathbf{x}}{\partial t^2} + \gamma\frac{\partial\mathbf{x}}{\partial t} + \delta_{\mathbf{x}}\mathcal{E} = \mathbf{f}. \tag{2}$$

During motion, the net external forces $\mathbf{f}(\mathrm{u}, t)$ balance dynamically against (i) the inertial force due to the mass density $\mu(\mathrm{u})$, (ii) the velocity-dependent damping force with damping density $\gamma(\mathrm{u})$, and (iii) the internal elastic force $\delta_{\mathbf{x}}\mathcal{E}$ which resists deformation. The partial differential equation (2) with appropriate conditions for $\mathbf{x}$ on the boundary of Ω and the initial configuration $\mathbf{x}(\mathrm{u}, 0)$ and velocity $\partial\mathbf{x}/\partial t|_{(\mathrm{u},0)}$ constitutes an initial boundary-value problem.

We express the elastic force, which acts to restore deformed bodies to their natural shapes, as a variational derivative $\delta_{\mathbf{x}}$ of a nonnegative deformation potential energy functional $\mathcal{E}(\mathbf{x})$. The energy $\mathcal{E}(\mathbf{x})$ should be zero for the model in its natural shape and it should grow as the total deformation of the model away from its natural shape increases. Furthermore, $\mathcal{E}(\mathbf{x})$ should be invariant to rigid motions of the model in Φ, since rigid motions impart no deformation. To define appropriate deformation energies we employ measures from differential geometry and the fundamental theorems regarding the rigidity of curves, surfaces, and solids in 3-space [7].

A reasonable functional for elastic bodies, which provides suitable invariance properties, is a Euclidean norm of the difference between the fundamental tensors of the deformed body and the fundamental tensors of the body in its natural shape. For example, the metric tensor $\mathbf{G}$ and curvature tensor $\mathbf{B}$ have entries

$$G_{ij} = \frac{\partial\mathbf{x}}{\partial u_i} \cdot \frac{\partial\mathbf{x}}{\partial u_j}; \qquad B_{ij}(\mathbf{x}(\mathrm{u}, t)) = \mathbf{n} \cdot \frac{\partial^2\mathbf{x}}{\partial u_i \partial u_j}, \tag{3}$$

where $\mathbf{n}(\mathrm{u}, t) = (\partial\mathbf{x}/\partial u_1 \times \partial\mathbf{x}/\partial u_2)/|\partial\mathbf{x}/\partial u_1 \times \partial\mathbf{x}/\partial u_2|$ gives the unit normal over the surface at time t. We distinguish with a superscript the body in its undeformed configuration,

$$\mathbf{x}^0(\mathrm{u}) = [x_1^0(\mathrm{u}), x_2^0(\mathrm{u}), x_3^0(\mathrm{u})]^{\mathsf{T}}, \tag{4}$$

as well as the fundamental tensors associated with this natural shape, $\mathbf{G}^0$, $\mathbf{B}^0$, etc. Then, for an elastically deformable solid we define the energy

$$\mathcal{E}(\mathbf{x}) = \int_\Omega |\mathbf{G} - \mathbf{G}^0|^2_{\mathbf{W}^1}\, du_1 du_2 du_3, \tag{5}$$

where the weighted matrix norm $|\cdot|_{\mathbf{W}^1}$ involves the weighting functions $w^1_{ij}(u_1, u_2, u_3)$. The analogous energy for a deformable surface is

$$\mathcal{E}(\mathbf{x}) = \int_\Omega |\mathbf{G} - \mathbf{G}^0|^2_{\mathbf{W}^1} + |\mathbf{B} - \mathbf{B}^0|^2_{\mathbf{W}^2}\, du_1 du_2, \tag{6}$$

where the weighted matrix norms $|\cdot|_{\mathbf{w}^1}$ and $|\cdot|_{\mathbf{w}^2}$, involve the weighting functions $w^1_{ij}(u_1, u_2)$ and $w^2_{ij}(u_1, u_2)$ respectively. For a deformable curve, we define the energy

$$\mathcal{E}(\mathbf{x}) = \int_\Omega w^1(s - s^0)^2 + w^2(\kappa - \kappa^0)^2 + w^3(\tau - \tau^0)^2 \, du, \tag{7}$$

where the scalar functions $s(\mathbf{x}(u,t))$, $\kappa(\mathbf{x}(u,t))$, and $\tau(\mathbf{x}(u,t))$ are the arc-length, curvature, and torsion respectively, and where $w^1(u)$, $w^2(u)$, and $w^3(u)$ are weighting functions.

The weighting functions in the above energies determine the properties of the simulated deformable material. In the curve energy (7), $w^1(u)$, $w^2(u)$, and $w^3(u)$ determine the resistance to stretching, bending, and torsion, respectively, along the curve. In the surface energy (6), $w^1_{ij}(u_1, u_2)$ control the surface's resistance to stretching and shearing, while $w^2_{ij}(u_1, u_2)$ control resistance to bending and twisting. In the energy for the solid (5), w^1_{ij} determine the resistance to stretching along u_1, u_2, and u_3, as well as shearing across planes perpendicular to these axes.

2.2. Hybrid Formulation.

Hybrid models include explicit deformable and rigid dynamics operating in concert. To formulate these models, we represent the positions of mass elements by

$$\mathbf{q}(\mathrm{u}, t) = \mathbf{r}(\mathrm{u}, t) + \mathbf{e}(\mathrm{u}, t), \tag{8}$$

the sum of a reference component $\mathbf{r}(\mathrm{u}, t) = [r_1(\mathrm{u}, t), r_2(\mathrm{u}, t), r_3(\mathrm{u}, t)]^\intercal$ and a deformation component $\mathbf{e}(\mathrm{u}, t) = [e_1(\mathrm{u}, t), e_2(\mathrm{u}, t), e_3(\mathrm{u}, t)]^\intercal$. Here, $\mathbf{r}$ and $\mathbf{e}$ are expressed relative to a body reference frame ϕ, with the subscripts 1, 2, and 3 denoting, respectively, the x, y, and z axes of ϕ.

The origin of ϕ coincides with the deformable body's center of mass

$$\mathbf{c}(t) = \int_\Omega \mu(\mathrm{u})\mathbf{x}(\mathrm{u}, t) \, du, \tag{9}$$

and ϕ is conveyed along with the body in accordance with the laws of rigid-body dynamics [6]. We express the linear and angular velocities of ϕ relative to the inertial frame Φ as

$$\mathbf{v}(t) = \frac{d\mathbf{c}}{dt}; \qquad \boldsymbol{\omega}(t) = \frac{d\boldsymbol{\theta}}{dt}, \tag{10}$$

where $d\boldsymbol{\theta}$ is a quantity whose magnitude equals the infinitesimal rotation angle and whose direction is along the instantaneous axis of rotation of ϕ relative to Φ. Then, the velocity of mass elements of the model relative to Φ, given their velocities $\partial\mathbf{e}(\mathrm{u},t)/\partial t$ relative to ϕ, is

$$\frac{\partial \mathbf{x}}{\partial t}(\mathrm{u},t) = \mathbf{v} + \boldsymbol{\omega} \times \mathbf{q} + \frac{\partial \mathbf{e}}{\partial t}. \tag{11}$$

Using Lagrangian mechanics, we transform the kinetic energy of the primal model (2) according to the above decomposition. Assuming small deformations, this yields the following three coupled differential equations for the unknown functions $\mathbf{v}$, $\boldsymbol{\omega}$, and $\mathbf{e}$ under the action of an applied force $\mathbf{f}(\mathrm{u},t)$:

$$m\frac{d\mathbf{v}}{dt} + \frac{d}{dt}\int_\Omega \mu\frac{\partial \mathbf{e}}{\partial t}\,d\mathrm{u} + \int_\Omega \gamma\frac{\partial \mathbf{x}}{\partial t}\,d\mathrm{u} = \mathbf{f}^{\mathbf{v}}; \tag{12a}$$

$$\frac{d}{dt}(\mathbf{I}\boldsymbol{\omega}) + \frac{d}{dt}\int_\Omega \mu\mathbf{q} \times \frac{\partial \mathbf{e}}{\partial t}\,d\mathrm{u} + \int_\Omega \gamma\mathbf{q} \times \frac{\partial \mathbf{x}}{\partial t}\,d\mathrm{u} = \mathbf{f}^{\boldsymbol{\omega}}; \tag{12b}$$

$$\mu\frac{\partial^2 \mathbf{e}}{\partial t^2} + \mu\frac{d\mathbf{v}}{dt} + \mu\boldsymbol{\omega} \times (\boldsymbol{\omega} \times \mathbf{q}) + 2\mu\boldsymbol{\omega} \times \frac{\partial \mathbf{e}}{\partial t} + \mu\frac{d\boldsymbol{\omega}}{dt} \times \mathbf{q} + \gamma\frac{\partial \mathbf{x}}{\partial t} + \delta_{\mathbf{e}}\mathcal{E} = \mathbf{f}. \tag{12c}$$

Here $m = \int_\Omega \mu\,d\mathrm{u}$ is the total mass of the body, and the inertia tensor $\mathbf{I}$ is given by

$$I_{ij} = \int_\Omega \mu(\delta_{ij}\mathbf{q}\cdot\mathbf{q} - q_i q_j)\,d\mathrm{u}, \tag{13}$$

where $\mathbf{q} = [q_1, q_2, q_3]^\mathsf{T}$ and δ_{ij} is the Kronecker delta. The applied force $\mathbf{f}(\mathrm{u},t)$ contributes to elastic deformation, as well as to a net translational force $\mathbf{f}^{\mathbf{v}}$ and a net torque $\mathbf{f}^{\boldsymbol{\omega}}$ on the center of mass:

$$\mathbf{f}^{\mathbf{v}}(t) = \int_\Omega \mathbf{f}\,d\mathrm{u}; \qquad \mathbf{f}^{\boldsymbol{\omega}}(t) = \int_\Omega \mathbf{q} \times \mathbf{f}\,d\mathrm{u}. \tag{14}$$

The ordinary differential equations (12a) and (12b) describe $\mathbf{v}$ and $\boldsymbol{\omega}$, the translational and rotational motion of the body's center of mass. The terms on the left hand sides of these equations pertain to the total moving mass of the body as if concentrated at $\mathbf{c}$, the total (vibrational) motion of the mass elements about the reference component $\mathbf{r}$, and the total damping of the moving mass elements. The partial differential equation (12c) describes the deformation $\mathbf{e}$ of the model away from $\mathbf{r}$ relative to ϕ. Each of its terms is a dynamic force per mass element: (i) the basic inertial force, (ii) the inertial force due to linear acceleration of ϕ, (iii) the centrifugal force due to the rotation of ϕ, (iv) the Coriolis force due to the velocity of the mass elements in ϕ, (v) the transverse force due to the angular acceleration of ϕ, (vi) the damping force, and (vii) the elastic restoring force due to deformation away from $\mathbf{r}$.

Once again, we represent the restoring force in (12c) as a variational derivative of an elastic potential energy functional $\mathcal{E}$. In the hybrid formulation, $\mathcal{E}$ need not be rigid-motion invariant, since it merely characterizes the deformational displacement $\mathbf{e}(\mathrm{u}, t)$ relative to the moving frame ϕ. This allows us to use linear restoring forces. We employ controlled-continuity spline energies [**9**] of the form

$$\mathcal{E}(\mathbf{e}) = \frac{1}{2} \int_{\Omega} \sum_{m=0}^{p} \sum_{|j|=m} \frac{m!}{j_1! \ldots j_d!} w_j \left|\partial_j^m \mathbf{e}\right|^2, \tag{15}$$

where $j = (j_1, \ldots, j_d)$ is a multi-index with $|j| = j_1 + \cdots + j_d$, and $\partial_j^m = \partial^m / \partial u_1^{j_1} \ldots \partial u_d^{j_d}$. Thus the energy density under the integral is a weighted sum of the magnitude of the deformation $\mathbf{e}$ and of its partial derivatives with respect to material coordinates. Generally, the order p of the highest partial derivative included in the sum determines the order of smoothness of the deformation.

As in the primal formulation, the weighting functions $w_j(\mathrm{u})$ in (15) control the properties of the deformable model over the material coordinates. In the case of a surface ($d = 2$) for instance, the function w_{00} penalizes the total magnitude of the deformation; w_{10} and w_{01} penalize the magnitude of its first partial derivatives; w_{20}, w_{11}, and w_{02} penalize the magnitude of its second partial derivatives; etc.

3. Inelastic Models.

Deformable behavior which does not obey the (Hookean) constitutive laws of classical elasticity is known as "inelastic". High polymer solids such as plasticine or modeling clay are strongly inelastic, and this makes them useful for molding complex shapes (e.g., for the design of automobile bodies). We have created inelastic models, a type of "computational modeling clay", by suitably extending our formulation of elastic models (see [**10**] for details).

Inelastic models incorporate three canonical genres of inelastic behavior: viscoelasticity, plasticity, and fracture. Viscoelastic behavior blends the characteristics of a viscous fluid together with elasticity. Whereas an elastic model, by definition, is one which has memory only of its reference shape, the current deformation of a viscoelastic model is a function of the entire history of applied force. Inelastic materials for which permanent deformations result from the mechanism of slip or atomic dislocation are known as plastic. Most metals, for instance, behave elastically only when the applied forces are small, after which they yield plastically, resulting in permanent dimensional changes. As materials are deformed beyond their structural limits, they eventually fracture, releasing excessive strain energy.

To include inelastic behavior into our primal formulation of elastic models, we incorporate a feedback process which evolves the reference tensor $\mathbf{G}^0$ (as well any higher-order reference tensors that may be present in $\mathcal{E}(\mathbf{x})$ for deformable surface or curve models) according to the model's instantaneous internal stresses or deformations. Furthermore, internal processes associated with plasticity and fracture dynamically adjust the material property functions; e.g., weighting functions $w_{ij}^1(\mathrm{u})$ in (5). In the hybrid formulation, we incorporate processes that update the reference component $\mathbf{r}$ and modify material properties according to applied force and instantaneous deformation $\mathbf{e}$. In effect, we allow $\mathbf{e}(\mathrm{u},t)$, as governed by (15), to play the role of a multidimensional elastic spring, while $\mathbf{r}(\mathrm{u},t)$ plays the role of a viscous dashpot; for instance, simple (Maxwell) viscoelasticity involves the feedback process $\dot{\mathbf{r}}(\mathrm{u},t) = (1/\eta(\mathrm{u}))\mathbf{e}(\mathrm{u},t)$.

Fractures are introduced into the elastic models using the distributed parameter functions. In the controlled-continuity spline energy (15), for instance, deformational discontinuities of order $0 \leq k < p$ are free to occur at any material point u_0 such that $w_j(\mathrm{u}_0) = 0$ for $|j| > k$ [**9**]. An internal fracture process automatically monitors stress and deformation distributions over Ω, and when local stress or deformation exceeds prescribed fracture limits, the process nullifies the w_j as necessary to introduce the appropriate discontinuities.

4. Thermoplastic Models.

Thermoplastic models [**11**] are deformable models capable of conducting heat into their interiors as soon as they come into contact with "hot" graphics objects in a simulated physical world. These models simulate thermoplastic materials which may be easily formed into desired shapes by pressure at relatively moderate temperatures, then made elastic or rigid in these shapes by cooling. Beyond thermoplasticity, the models feature melting and fluid-like behavior in the molten state.

In addition to their nonrigid dynamics, governed by the Lagrangian equations of motion (2), the models transfer heat through their interiors according to the heat equation for nonhomogeneous, nonisotropic conductive media. In the case of solids, the heat equation which governs the temperature distribution $\theta(\mathrm{u}, t)$ may be written as

$$\frac{\partial}{\partial t}(\mu\sigma\theta) - \nabla \cdot (\mathbf{C}\nabla\theta) = q, \tag{16}$$

where $\mu(\mathrm{u})$ is the mass density, $\sigma(\mathrm{u})$ is the specific heat, $\nabla = [\partial/\partial u_1, \partial/\partial u_2, \partial/\partial u_3]^\intercal$ is the gradient operator in material coordinates, $\mathbf{C}$ is the (symmetric) thermal conductivity matrix of the simulated material, and $q(\mathrm{u}, t)$ is the rate of heat generation (or loss) per unit volume.

The thermoplastic (softening) region in our models is defined by a linear force-versus-deformation relationship minus a component proportional to the temperature (Duhamel–Neumann thermoelasticity law). To implement this law, we establish a relationship between the temperature θ and the material property functions $\mathbf{w}_i$ in, e.g., (5). As the temperature increases, the stiffness of a thermoplastic model decreases. Eventually the model fractures into "fluid" particles when its temperature exceeds the melting point.

In its molten state the deformable model involves a many-body simulation in which "fluid" particles that have broken free interact through long-range attraction forces and short-range repulsion forces. Suppose particle i has mass m_i and is located at $\mathbf{x}_i(t)$, while particle j has mass m_j and is located at $\mathbf{x}_j(t)$ at time t. Let $r_{ij}(t) = |\mathbf{x}_i - \mathbf{x}_j|$ be the separation of the two particles. We define the force on particle i exerted by particle j as

$$\mathbf{g}_{ij}(t) = m_i m_j(\mathbf{x}_i - \mathbf{x}_j)\left(-\frac{\alpha}{(r_{ij}+\zeta)^a} + \frac{\beta}{(r_{ij})^b}\right), \tag{17}$$

where α and β are nonnegative parameters that determine the strength of the attraction and repulsion components of the force, and ζ is a positive measure of how close the particles are allowed to be [**12**].

5. Remarks.

Forces and constraints: Applying external forces to deformable models yields realistic animations. The net external force $\mathbf{f}(\mathrm{u}, t)$ in (2) or (12c) is the sum of the individual external forces. A plethora of interesting force functions is conceivable. We have implemented simple instances of gravitational forces, aerodynamic forces, and repulsive forces due to collisions with impenetrable objects, friction forces, and soft constraint forces (see [**3,4**] for details). The latter include interactive constraints applied by the user. Techniques for imposing hard constraints on deformable models have been devised based on optimization theory [**13**]. Using forces and constraints, we are able to synthesize complex motions arising from the interaction of deformable models with simulated physical worlds.

Numerical implementation: To create animation with deformable models, we simulate numerically their differential equations of motion. The first step is to discretize the partial differential equations (2) or (12c) in material coordinates. We use finite-difference or finite-element approximations on a discrete mesh of nodes [**14**]. To simulate the dynamics of deformable models, the resulting large system of simultaneous ordinary differential equations is then integrated through time using a semi-implicit time integration procedure. After each time step (or every few time steps) in the simulation, we render the models' state data to create successive frames of an animation sequence. In essence, the evolving deformation yields a recursive sequence of (dynamic) equilibrium problems, each requiring solution of a sparse linear system, whose dimensionality is proportional to the number of nodes comprising the discrete deformable model. It is crucial to choose applicable numerical solution methods judiciously in order to achieve efficiency. We have used a variety of methods, depending on the size of the problem. These include efficient direct (Choleski) methods, iterative (successive over-relaxation, conjugate gradient) methods, hybrid (alternating direction implicit) methods, and multigrid methods. The mathematical details of our implementation for the case of elastic surfaces are given in [**3**]. Details for the case of thermoelastic models are found in [**11**].

Vision: The inverse problem to computer graphics is computer vision [**9**]. Whereas graphics involves the synthesis of images from models of objects, vision is concerned with the analysis of images in order to infer models of imaged objects; i.e., reconstruction of 3-dimensional models from 2-dimensional intensity data. A powerful approach to solving the reconstruction problems of computer vision is to use deformable models. We immerse deformable models in ambient force fields computed from image data. The force fields treat the models as "modeling clay", molding them into shapes that are consistent with imaged objects of interest. When objects move, their images change and the force fields are perturbed, thereby moving the models nonrigidly to maintain consistency over time. See [**15**] for examples of inverse modeling using deformable curve and surface primitives.

Acknowledgement.

John Platt, Andrew Witkin, and especially Kurt Fleischer made significant contributions to the work described in this paper, which was carried out at Schlumberger Palo Alto Research and at the Schlumberger Laboratory for Computer Science. Kurt Fleischer rendered the images using the modeling testbed system described in [**16**].

REFERENCES

1. Mortensen, M. E., "Geometric Modeling," Wiley, New York, NY, 1985.
2. Lassiter, J., *Principles of traditional animation applied to 3D computer animation*, Computer Graphics **21**, 4 (1987), 35–44.
3. Terzopoulos, D. and Fleischer, K., *Deformable models*, The Visual Computer **4**, 6 (1988), 306–331.
4. Terzopoulos, D.; Platt, J.; Barr, A. and Fleischer, K., *Elastically deformable models*, Computer Graphics **21**, 4 (1987), 205–214.
5. Terzopoulos, D. and Witkin, A., *Physically-based models with rigid and deformable components*, IEEE Computer Graphics and Applications **8**, 6 (1988), 41–51.
6. Goldstein, H., "Classical Mechanics, 2nd Ed.," Addison–Wesley, Reading, MA, 1980.
7. do Carmo, M. P., "Differential Geometry of Curves and Surfaces," Prentice–Hall, Englewood Cliffs, NJ, 1974.
8. Faux, J. D. and Pratt, M. J., "Computational Geometry for Design and Manufacture," Halstead Press, Horwood, NY, 1981.
9. Terzopoulos, D., *Regularization of inverse visual problems involving discontinuities*, IEEE Trans. Pattern Analysis and Machine Intelligence **PAMI-8** (1986), 413–424.
10. Terzopoulos, D. and Fleischer, K., *Modeling inelastic deformation: Viscoelasticity, plasticity, fracture*, Computer Graphics **22**, 4 (1988), 269–278.
11. Terzopoulos, D.; Platt, J. and Fleischer, K., *Heating and melting deformable models*, Proc. Graphics Interface '89, London, Canada, June 1989, 219–226.
12. Greenspan, D., "Discrete Models," Addison–Wesley, Reading, MA, 1973.
13. Platt, J. and Barr, A., *Constraint methods for flexible models*, Computer Graphics **22**, 4 (1988), 279–288.
14. Kardestuncer, H. and Norrie, D. H., (ed.), "Finite Element Handbook," McGraw–Hill, New York, NY, 1987.
15. Terzopoulos, D.; Witkin, A. and Kass, M., *Constraints on deformable models: Recovering 3D shape and nonrigid motion*, Artificial Intelligence **35** (1988), 91–123.
16. Fleischer, K. and Witkin, A., *A modeling testbed*, Proc. Graphics Interface '88, Edmonton, Canada, June 1988.

University of Toronto, Department of Computer Science, Toronto, Canada, M5S 1A4; Canadian Institute for Advanced Research; and Schlumberger Laboratory for Computer Science, P.O. Box 200015, Austin, TX 78720

Periodic Area Minimizing Surfaces in Microstructural Science

EDWIN L. THOMAS, DAVID M. ANDERSON, DAVID C. MARTIN,
JAMES T. HOFFMAN, AND DAVID HOFFMAN

ABSTRACT

An A/B block copolymer consists of two macromolecules bonded together. In forming an equilibrium structure, such a material may separate into distinct phases, creating domains of component A and component B. A dominant factor in the determination of the domain morphology is area-minimization of the intermaterial surface, subject to fixed volume fraction. Surfaces that satisfy this mathematical condition are said to have constant mean curvature. The geometry of such surfaces strongly influences physical properties of the material, and they have been proposed as structural models in a variety of physical and biological systems. We have discovered domain structures in phase-separated diblock copolymers that closely approximate periodic constant mean curvature surfaces. Transmission electron microscopy and computer simulation are used to deduce the three-dimensional microstructure by comparison of tilt series with two-dimensional image projection simulations of 3-D mathematical models. Three structures are discussed: the first of which is the double diamond microdomain morphology associated to a newly discovered family of triply periodic constant mean curvature surfaces. Second, a doubly periodic boundary between lamellar microdomains, corresponding to a classically known surface (called Scherk's First Surface), is described. Finally, we show a lamellar morphology in thin films that is apparently related to a new family of periodic surfaces.

Some References

Work on Polymers

1. E. L. Thomas et al. Nature 334, 598–601 (1988)
2. D. M. Anderson and E. L. Thomas. Macromolecules 21, 3221 (1988)

Work on Surfactant Systems

3. L. E. Scriven. Nature 263, 123–124 (1976)
4. D. M. Anderson, et al. Proc R. Soc. Phil Trans (in press)
5. J. Charvolin. J. Physique 46, 173–183 (1985)

Work for the Mathematically Inclined with Access to Supercomputing

6. M. J. Callahan, et al. Comm ACM31, 648–661 (1988)
7. D. M. Anderson. PhD Thesis, University of Minnesota (1985)

Department of Polymer Science, University of Massachusetts, Amherst, MA 01003 and
Department of Mathematics, University of Massachusetts, Amherset, MA 01003

Types of Instability for the Trapped Drop Problem with Equal Contact Angles

THOMAS I. VOGEL

This paper continues the study of the stability of a liquid drop forming a bridge between two parallel planes in the absence of gravity (Figure 1) (see [2], [3], [4]). It is important to realize that the curves of contact between the free surface of the drop and the planar plates are assumed to be free. In other words, a perturbation of the drop which changes the wetted regions on the planes is admissible. The parameters which are determined physically are the volume of the drop and the angles of contact between the free surface and the two fixed planes.

I will summarize the results of [3] and [4] necessary to the present paper. If a stable drop is formed, it must be rotationally symmetric, with each cross section of the drop parallel to the fixed planes being a disk, and therefore the free surface may be described by $f(x)$, $x \in [0, h]$ where h is the separation of the planes and $f(x_0)$ is the radius of the trace of the free surface on the plane $x = x_0$. This function must satisfy

$$(1)\qquad \begin{aligned} \mathcal{M}(f) &= \frac{1}{2}\left(\frac{f''}{(1+(f')^2)^{\frac{3}{2}}} - \frac{1}{f(1+(f')^2)^{\frac{1}{2}}}\right) - H \\ f'(0) &= -\cot\gamma_1 \\ f'(h) &= \cot\gamma_2 \end{aligned}$$

for some constant H, where $\mathcal{M}$ is the rotationally symmetric mean curvature operator, and γ_1 and γ_2 are the contact angles with Π_1 and Π_2. In [4] it is shown that if f solves (1) and if the quadratic form

$$\beta(\psi) \equiv \int_0^h \frac{f(\psi')^2}{(1+(f')^2)^{\frac{3}{2}}} - \frac{\psi^2}{f(1+(f')^2)^{\frac{1}{2}}}dx$$

is positive definite on the space of functions

$$f^{\perp} = \{\psi \in C^1 : \int_0^h \psi f dx = 0\},$$

then f is a local minimum for the relevant energy functional. If $\beta(\psi) < 0$ for any $\psi \in f^{\perp}$, then f is not a local minimum. ψ may be thought of as an infinitesimally volume conserving perturbation of f, and $\beta(\psi)$ the second variation. In the same paper, it is shown that if the following two conditions are satisfied, then $\beta(\psi)$ is positive definite on $f^{\perp}$.

1) The Sturm–Liouville problem

$$L(\psi) \equiv -\left(\frac{f\psi'}{(1+(f')^2)^{\frac{3}{2}}}\right)' - \frac{\psi}{f(1+(f')^2)^{\frac{1}{2}}} = \lambda\psi \tag{2}$$
$$\psi'(0) = \psi'(h) = 0$$

has precisely one negative eigenvalue, and

2) $f(x) \equiv f(x;\varepsilon_0)$ may be embedded in a smoothly parametrized family $f(x;\varepsilon)$ of solutions to equation (1) with $H = H(\varepsilon)$, and γ_1 and γ_2 fixed; with

$$H'(\varepsilon_0)V'(\varepsilon_0) > 0,$$

where $V(\varepsilon)$ is the volume of the drop corresponding to $f(x;\varepsilon)$.

If $L(\psi)$ has 2 or more negative eigenvalues, then the drop is unstable, and if $H'(\varepsilon_0)V'(\varepsilon_0) < 0$, the drop is unstable.

Because of the importance of V and H in determining stability, we will need to consider the image of families of drops under the $H - V$ map. Figure 2, for example, shows the image of zero-inflection, one-inflection, and two-inflection drops with $\gamma_1 = \gamma_2 = 60°$ in the dimensionless $H - V$ plane. Each point on one of the curves represents a drop with the appropriate volume and mean curvature.

In [**4**] it was shown that $\lambda_1 > 0 > \lambda_0$ for any solution to (1) whose second derivative is bounded from zero. Thus, interior to the family of inflectionless profiles for some fixed γ_1 and γ_2, the only way an instability can develop is by condition 2 failing. If this occurs, the point in the $H - V$ plane corresponding to the profile curve at which $\frac{dV}{dH}$ changes sign will be called a point of instability of type 2.

The other way a family of drops can become unstable is by condition 1 failing, i.e. λ_1 passing through zero (while condition 2 remains true). The point in the $H - V$ plane corresponding to this is a point of instability of type 1.

We now consider the case $\gamma_1 = \gamma_2 \equiv \gamma$. Both types of instabilities have been observed numerically in the family of profiles without inflections (see [**2**] for more complete numerical results), although a type 1 instability could only occur if $f'' = 0$ at an endpoint. In Figure 2, for $\gamma_1 = \gamma_2 = 60°$, the point of instability is of type 1, and corresponds to the bifurcation which occurs when an inflection appears on the boundary. Numerically, in Figure 2, the family of two-inflection profiles and the double family of one-inflection profiles are both unstable due to λ_1 being negative. The fact that, for $\gamma_1 = \gamma_2$ the two-inflection profiles have at least two negative eigenvalues for (2) will be proven in [**5**]. In the present paper we will assume this.

An instability of type 2 is illustrated in Figure 3 for $\gamma_1 = \gamma_2 = 20°$ This occurs interior to the family of profiles without inflections. Numerically, it appears that there is an angle $\gamma_0 \approx 31.14°$ such that for $\gamma_1 = \gamma_2 > \gamma_0$, the family of inflectionless profiles will exhibit a type 1 instability and for $\gamma_1 = \gamma_2 < \gamma_0$, the family of inflectionless profiles will exhibit a type 2 instability ([**2**]). (If $\gamma_1 \neq \gamma_2$, no instability of type 1 was ever observed, i.e. $\frac{dV}{dH}$ had always changed sign before the bifurcation).

The purpose of the present paper is to predict a qualitative difference in the behavior at these two types of instabilities (for $\gamma_1 = \gamma_2$). The following will be shown. At a type 1 instability, the drop remains stable if only perturbations symmetric to the plane $x = \frac{h}{2}$ are considered and only becomes unstable if non-symmetric perturbations are allowed. At a type 2 instability, the drop becomes unstable with respect to a perturbation symmetric across $x = \frac{h}{2}$.

LEMMA. *If $f(x)$ is symmetric with respect to $x = \frac{h}{2}$, then the eigenfunctions of* (2) *corresponding to λ_n with n even satisfy $\psi(h - x) = \psi(x)$, and the eigenfunctions of* (2) *corresponding to λ_n with n odd satisfy $\psi(h-x) = -\psi(x)$.*

PROOF: Let $\psi(x)$ be an eigenfunction of (2). Since the coefficients of (2) are symmetric with respect to $x = \frac{h}{2}$, $\psi(h-x)$ must also be an eigenfunction of (2) corresponding to the same eigenvalue. The eigenvalues of a regular Sturm–Liouville problem are simple [**1**], so $\psi(h - x) = c\psi(x)$. It is easy to see that since $\psi(x)$ is not identically zero, $c = \pm 1$.

If $\psi(h-x) = \psi(x)$, then $\psi'(\frac{h}{2}) = 0$. This implies that $\psi(\frac{h}{2}) \neq 0$. Indeed, if this were not the case, ψ would be identically zero by the uniqueness theorem for solutions of O.D.E.'s. Therefore ψ will have an even number of zeros and hence will correspond to λ_n with n even ([**1**]). If $\psi(h - x) = -\psi(x)$, then $\psi(\frac{h}{2}) = 0$ and $\psi'(\frac{h}{2}) \neq 0$. ψ will have an odd number of zeros, and will correspond to a λ_n with n odd. □

We are now in a position to compare the trapped drop problem for $\gamma_1 = \gamma_2 = \gamma$ with that for $\gamma_1 = \gamma, \gamma_2 = 90°$. Given a solution f_1 to (1) with

$\gamma_1 = \gamma_2 = \gamma$ which is symmetric across $x = \frac{h}{2}$, a solution f_2 to (1) with $\gamma_1 = \gamma$, $\gamma_2 = 90°$ may be obtained by taking f_1 on $[0, \frac{h}{2}]$ and scaling by $f_2(x) = 2f_1(\frac{x}{2})$. The drop whose profile is f_2 will have four times the volume and half the mean curvature of the drop whose profile is f_1. Moreover, by the preceding lemma, if the eigenvalues of (2) for f_1 are $\lambda_0, \lambda_1, \lambda_2, \ldots$, then the eigenvalues for f_2 are $\frac{\lambda_0}{2}, \frac{\lambda_2}{2}, \frac{\lambda_4}{2} \ldots$ with the factor of 2 coming from the scaling. This is because the even eigenfunctions for f_2 satisfy $\psi'(\frac{h}{2}) = 0$ and the odd ones do not.

Let G be the subset of $f^\perp$ consisting of all $\phi \in f^\perp$ which satisfy $\int_0^{\frac{h}{2}} \phi f \, dx = 0$ (and of course $\int_{\frac{h}{2}}^h \phi f \, dx = 0$), and let $f_s^\perp$ be the subset of G consisting of all $\phi \in G$ with $\phi(h - x) = \phi(x)$. G may be thought of as the set of infinitesimally volume preserving perturbations which cause no fluid flow across the plane $x = \frac{h}{2}$, and $f_s^\perp$ consists of infinitesimally volume preserving perturbations which are symmetric with respect to $\frac{h}{2}$.

THEOREM. *Suppose that $p_0 = (H_0, V_0)$ is a point of instability of the family of inflectionless profiles for the problem $\gamma_1 = \gamma_2 = \gamma$. If p_0 is of type 1, then it is no longer a point of instability of we restrict the perturbations to lie in G and the drops to be symmetric across $x = \frac{h}{2}$. If p_0 is of type 2, then it will remain a point of instability if we restrict ourselves to perturbations in $f_s^\perp$.*

PROOF: Suppose that p_0 is an instability of type 1. Then p_0 is the bifurcation occuring when a inflection appears on the boundary. In going from the zero-inflection family to the two-inflection family, λ_1 crosses zero. Restricting ourselves to perturbations in G and drops symmetric with respect to $\frac{h}{2}$ is equivalent (after scaling) to two copies of the problem $\gamma_1 = \gamma$, $\gamma_2 = 90°$. For this problem, $(4V_0, \frac{H_0}{2})$ (the point corresponding to p_0) is not a point of instability, since $\frac{dV}{dH}|_{p_0} > 0$ (the instability is of type 1), and $\frac{\lambda_0}{2} < 0 < \frac{\lambda_2}{2} < \cdots$ at $(4V_0, \frac{H_0}{2})$. Thus restricting the perturbations and drops as above removes the instability.

Now suppose p_0 is an instability of type 2, so that $\frac{dV}{dH}$ changes sign at p_0. Restricting perturbations to $f_s^\perp$ is equivalent to one copy of the problem $\gamma_1 = \gamma$, $\gamma_2 = 90°$. In this case, however, $(4V_0, \frac{H_0}{2})$ is a point of instability for $\gamma_1 = \gamma$, $\gamma_2 = 90°$, since $\frac{dV}{dH}$ along the $H - V$ curve of $\gamma_1 = \gamma_2$, $\gamma = 90°$ must also change sign. (The $H - V$ curve of $\gamma_1 = \gamma_2$, $\gamma = 90°$ is obtained by scaling the parts of the $H - V$ curve of $\gamma_1 = \gamma_2 = \gamma$ which represent drops symmetric with respect to $x = \frac{h}{2}$). Thus the original point of instability remains even if we restrict perturbations to $f_s^\perp$. □

Thus, when a type 1 instability develops as the volume decreases, a perturbation to which the drop becomes unstable cannot be symmetric with respect to the plane $x = \frac{h}{2}$, and must in fact cause a net fluid flow across that plane. Intuitively, it's reasonable to expect that after the drop breaks, the drops on each plane (which will be spherical caps) will be of different volumes. However, a proof of this is beyond the scope of this paper, since once the drop becomes unstable, dynamic effects must be taken into account. If instead an instability of type 2 is encountered, it seems likely that the break is symmetric, since the corresponding drop with $\gamma_1 = \gamma_2$, $\gamma = 90°$ is also unstable. Again, a proof of this is beyond the scope of this paper.

Acknowledgements. This work was partially supported by NSF grant DMS-8801515. In addition, some of the research was supported by the Applied Mathematical Sciences subprogram of the Office of Energy Research, U.S. Department of Energy, under contract DE-AC03-76SF00098 at the Lawrence Berkeley Laboratory.

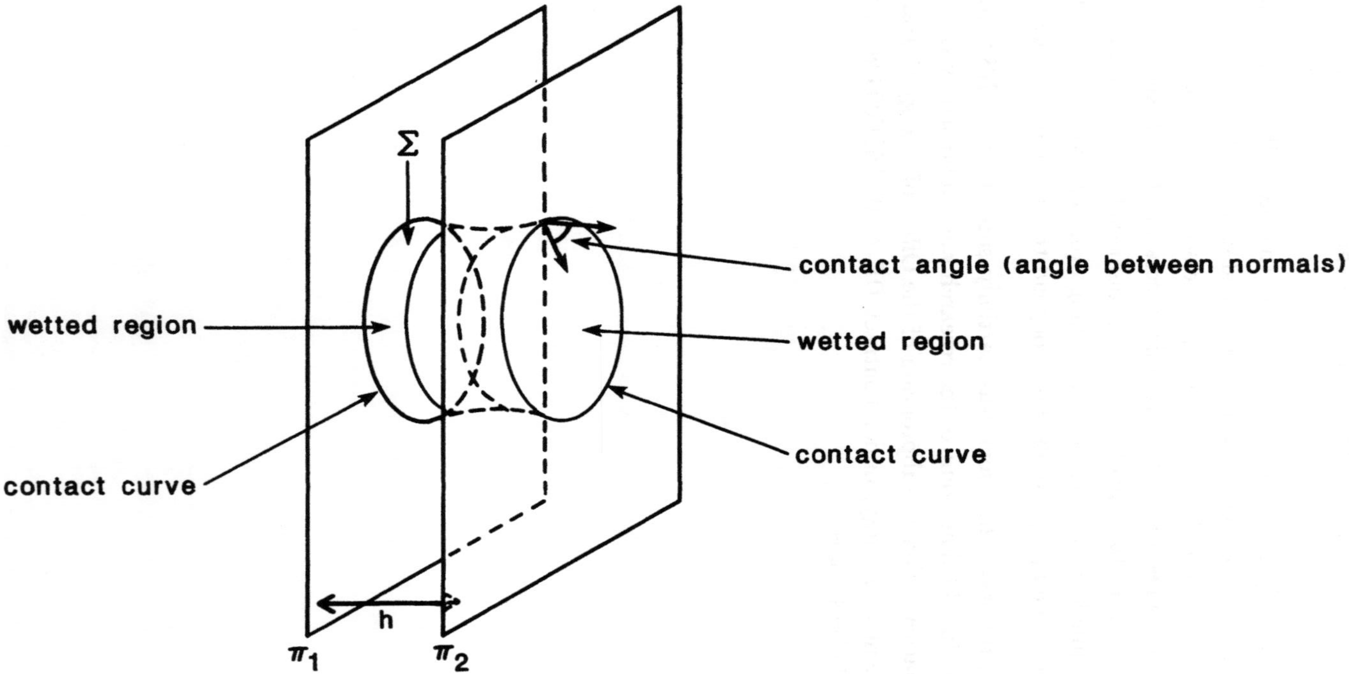

Figure 1: Liquid bridge between two parallel planes with separation h

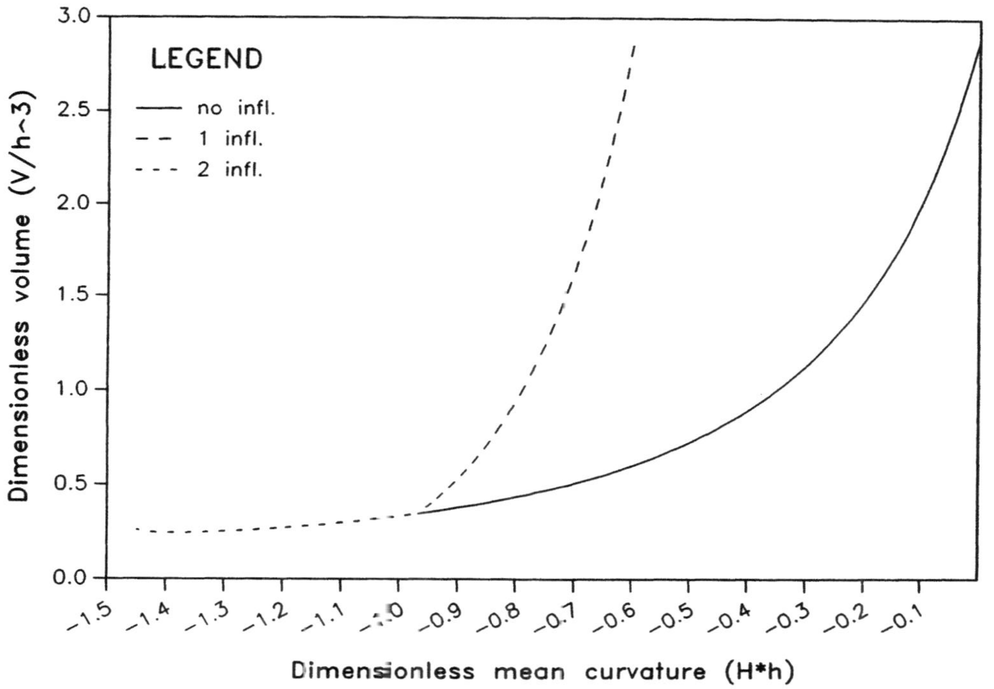

Figure 2: V vs. H for $\gamma_1 = \gamma_2 = 60°$

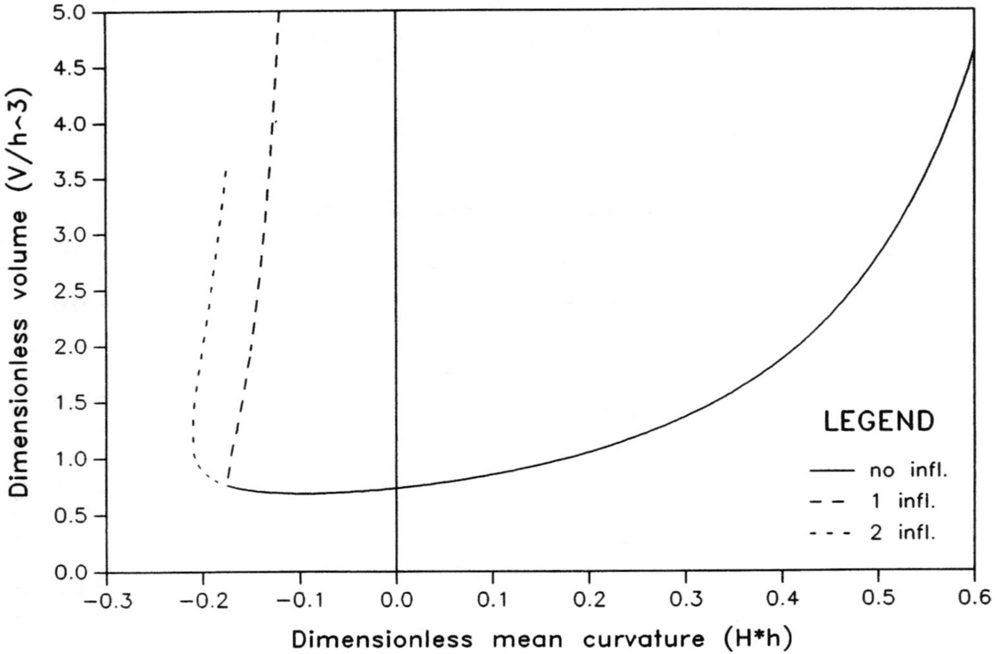

Figure 3: V vs. H for $\gamma_1 = \gamma_2 = 20°$

References

1. Birkhoff, G. and Rota, G. C., "Ordinary Differential Equations," 3rd ed, John Wiley & Sons, New York, 1978.
2. Vogel, T. I., *Numerical results on trapped drop stability*, to appear.
3. ————, *Stability of a liquid drop trapped between two parallel planes*, SIAM J. Appl. Math., Vol. 47, No. 3 (1987), 516–525.
4. ————, *Stability of a liquid drop trapped between two parallel planes* II*: general contact angles*, SIAM J. Appl. Math., Vol. 49, No. 4 (1989), 1009–1028.
5. ————, *Stability of a liquid drop trapped between two parallel planes* III, in preparation.

Department of Mathematics, Texas A&M University, College Station, TX 77843

MIX
Papier aus verantwortungsvollen Quellen
Paper from responsible sources
FSC® C105338

If you have any concerns about our products,
you can contact us on
ProductSafety@springernature.com

In case Publisher is established outside the EU,
the EU authorized representative is:
Springer Nature Customer Service Center GmbH
Europaplatz 3, 69115 Heidelberg, Germany

Printed by Libri Plureos GmbH
in Hamburg, Germany